KT-211-053

This book now belongs to you

It's part of the Guardian
and Observer Book Swap.

Thousands of books have been left
all over the country, you're lucky
enough to have found one.

Visit **guardian.co.uk/bookswap**
to find out more.

Here you can upload a photo of
where you found this book.

And once you've read it, you can
rate it and review it there too.

Then why not pass it on to
someone else?

🐦 #guardianbookswap

BOOKS SEASON

theguardian TheObserver

BRIAN CLEGG

INFLIGHT SCIENCE

A GUIDE TO THE WORLD FROM YOUR AIRPLANE WINDOW

ICON BOOKS

Published in the UK in 2011 by
Icon Books Ltd, Omnibus Business Centre,
39–41 North Road, London N7 9DP
email: info@iconbooks.co.uk
www.iconbooks.co.uk

Sold in the UK, Europe, South Africa and Asia
by Faber & Faber Ltd, Bloomsbury House,
74–77 Great Russell Street,
London WC1B 3DA or their agents

Distributed in the UK, Europe, South Africa and Asia
by TBS Ltd, TBS Distribution Centre, Colchester Road,
Frating Green, Colchester CO7 7DW

Published in Australia in 2011
by Allen & Unwin Pty Ltd,
PO Box 8500, 83 Alexander Street,
Crows Nest, NSW 2065

Distributed in Canada by
Penguin Books Canada,
90 Eglinton Avenue East, Suite 700,
Toronto, Ontario M4P 2YE

This edition published in the USA in 2011 by Totem Books
Inquiries to: Icon Books Ltd, Omnibus Business Centre,
39–41 North Road, London N7 9DP, UK

Distributed to the trade in the USA
by Consortium Book Sales and Distribution
The Keg House, 34 Thirteenth Avenue NE, Suite 101
Minneapolis, Minnesota 55413-1007

ISBN: 978-184831-241-8

Typeset in Melior by Marie Doherty

Printed and bound in the UK by
CPI Mackays, Chatham, Kent ME5 8TD

Contents

List of illustrations

About the author

Brian Clegg is a science writer (website: www.brianclegg.net). He runs www.popularscience.co.uk, and his most recent book was *Armageddon Science* (St Martin's Press, 2010).

Disclaimer

The experiments in this book are designed to be safe, and many of them can be done on board an aircraft. Those that are better carried out at home are clearly indicated. When carrying out any experiments in the air, make sure that you don't disturb other passengers or distract the cabin crew. Any experiments that could cause damage, danger or disturbance are clearly marked as **not to be performed** and are theoretical examples only. The publisher accepts no responsibility for any damage, injury or loss arising from any of the experiments contained in this book, theoretical or otherwise.

For Gillian, Chelsea and Rebecca

At the Airport

Terminal boredom

You're sitting in the terminal, waiting for the flight. A whole mix of conflicting emotions could be vying for attention: boredom, excitement and fear included. Boredom often wins. Flying may be the quickest way to get to a distant destination, but it includes a lot of waiting around.

Even if you're a seasoned traveller, though, there's something special about taking to the air, an excitement that's often triggered by the scent of kerosene on the tarmac, or the sound of an aircraft engine starting up. And there's an element of fear – because however much you enjoy flying, there's something highly unnatural about being suspended in a metal and plastic tube seven miles up, with only science and technology to keep you alive.

If you don't like flying (and I don't), a little science might help by providing some very reassuring statistics. The risk of being killed in a plane crash in any particular year is 1 in 125 million passenger journeys. This makes it three times safer on any particular journey than travelling by train – and when did you ever worry about that? The equivalent risk for a car is 1 in 10 million – twelve times as dangerous. You're more likely to have a fatal accident during six hours spent in the workplace than you are during six hours on a plane. There's only so much reassurance you can get from statistics – but flying is incredibly safe.

Our focus will be on what you see and experience on board an aircraft, but it's quite possible that boredom will kick in as you wait in the terminal. You can only do so many trips round the duty-free shops, or drink so many coffees. So let's take a brief look at some of the extreme technology you might encounter on the ground before taking to the air.

An airport divided

Airports have a strict divide between groundside and airside. To get from one to the other, particularly when flying internationally, you will face a barrage of technology aimed at identifying you and checking that you aren't carrying anything dangerous. If airlines were permitted, they would also weigh you as you pass through (this was done in the early days of flight). Plane loading is very sensitive to weight and airlines have to rely on average weights to know how much load the passengers are contributing.

Making such an estimate has, at least once, caused problems. The plane, taking off from a German airport, struggled to get away from the runway and only just managed to claw its way into the air. It later turned out that there was a coin fair on in the city, and many of the passengers were coin dealers with their pockets crammed with new acquisitions, because they didn't want to risk their new purchases being stolen from the hold baggage. All this unexpected spare change pushed the passengers' weight well above the expected average. Added up over the entire aircraft, there was so much extra load that the plane didn't respond as the pilots expected it to, causing a few worrying moments on take-off.

Bag check

Your first encounter with interesting technology is likely to be the security scanners. Your hand baggage is put on a conveyor belt that carries it through a powerful X-ray machine. That name 'X-ray' is not because of some special scientific naming convention, it's just that when discoverer Wilhelm Roentgen first came across rays that would pass through solid objects he called them X-rays (or rather *X-Strahlen*) to show that they were unknown and mysterious. They were officially renamed Roentgen rays, but everyone liked Roentgen's original nickname for them, and it stuck.

In reality, X-rays aren't particularly mysterious – they are nothing more or less than light, but light of a colour that is far outside the spectrum that we can see. All light is 'electromagnetic radiation', a special interaction between electricity and magnetism that comes in a huge range of 'colours'. As well as visible light there is radio, microwaves, infra-red, ultra-violet, X-rays and gamma rays – all exactly the same kind of stuff but with varying amounts of energy (see illustration overleaf). We now know that light is made up of tiny particles called photons (more on these later). X-rays consist of much higher-energy photons than visible light. If you prefer to think of light as a wave, as it was probably described to you at school, then X-ray waves have a shorter wavelength (the distance in which the wave makes a complete wiggle) than visible light.

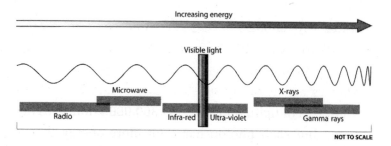

1. The electromagnetic spectrum: visible light forms a small segment near the middle.

When ordinary light hits an object like a suitcase that isn't transparent, the photons of light are absorbed. This happens because the energy in the photon is sucked up by one of the particles that make up the suitcase. Every object we see around us is made up of atoms, and each atom consists of a very small central part, the nucleus, which contains over 99 per cent of its weight, surrounded by a fuzz of tiny particles called electrons. When a photon of light meets an electron, the electron can consume the energy in the photon. This leaves the electron buzzing around with more energy than it started with.

This process of an electron absorbing or giving off the energy of a photon of light is called a quantum leap, a term that has come to mean a large, significant change, even though a real quantum leap is an absolutely tiny difference.

Once the electron has absorbed the energy of the photon, it's as if it were teetering on top of a high wall. Before long, that extra energy shoots back out in the form of a new photon, and the electron drops back to having less energy. We don't know in which direction a particular photon will shoot off, but over time some will head towards your eyes. It's these photons, pumped out by the electrons in an object, that allow you to see it.

X-rays are made up of photons too, and as with all other forms of light, they travel at 300,000 kilometres per second, but each photon has a lot more energy than a photon of ordinary light, enabling it to smash past the electrons in an object's atoms with much less interaction. This means that X-rays can penetrate many substances that stop ordinary light dead.

In the process of battering through matter, the X-rays can cause damage to the molecules (molecules are just collections of atoms that are joined together) that make up an object. Each cell in the human body contains huge molecules of DNA that contain the instructions for how the cell should behave. If these molecules, or other important chemicals in the cell, are damaged by the impact of X-rays, the changes can increase the chances of cancer forming. This is why medical X-rays have to be used with care, keeping doses to a minimum. Before the 1960s this wasn't appreciated. You still saw X-ray devices in shoe shops, for example, where you could peer through and see your toe bones wriggling around inside the shoe.

Inanimate objects are less susceptible to damage (though photographic film can be fogged), so baggage X-ray machines are considerably more powerful than most medical X-rays. Those big scanners you now find in airports use a wide band of X-rays, some more powerful than others. After passing through your bag and its contents, the X-rays reach detectors, working on a similar principle to a digital camera. There are two sets of sensors, one behind the other, separated by a metal shield. The weaker X-rays are stopped by the metal shield, so register only on the front detectors, but the more powerful X-rays blast on through the shield, so are spotted by both.

This distinction between the two strengths of X-ray is used to produce different-coloured images on the operator's screen. This way the picture will distinguish between 'soft' matter like plants, plastic or explosives – which are usually coloured orange by the scanner – and less penetrable matter, which will let only the more powerful X-rays through – typically coloured green. The result is to give more depth to the image and to distinguish at a glance between the different types of material within your baggage.

Testing the air

It's also possible that your bags will be subjected to a sniffer, hunting down explosives by their smell. Like many substances, explosives are to some extent volatile. This means that some of the molecules making up the chemicals within the explosive evaporate at room temperature and waft into the air. Molecules in solids and liquids are always bouncing around, and some bounce with more energy than the rest, managing to escape. This is the process that makes it possible for us to smell anything, whether we're sniffing the bouquet of a glass of wine, or appreciating the tempting odour of baking bread. It's also why a pool of water will eventually evaporate, even at room temperature.

Sometimes the sniffer will be a dog. Arguably, the dog is the oldest piece of highly developed technology still in active use. It might seem bizarre to call a dog 'technology'. Yet dogs have been consciously moulded into distinct breeds with specific functions in mind. They were the first autonomous technology – they function on their own, as opposed to a similarly ancient device like a hand axe that had to be powered by a human being. Now we have dogs that provide a wide range of functions, from guide dogs and sheepdogs to the owners of the extremely sensitive noses that can pick out the presence of explosives.

Of course the production of this remarkable piece of technology didn't originate with the intention of creating such a flexible helper. The chances are it all started by accident, when wolves began to hang around human camps. Although wolves don't deserve a lot of the bad press they get – they rarely attack human beings, for instance – they would have been irritating scavengers that early man had to make an effort to see off, to stop them stealing the remains of hunted animals.

It's easy to imagine those first, tentative steps away from the wolf's role as enemy. Perhaps it was a cold winter, and a wolf crept close to a fire to keep warm. Maybe while it was there some other predator attacked the camp – the wolf, ever the pack animal, jumped to the defence of the humans, fighting alongside them. It was rewarded with a gift of meat. Wolf cubs that were more docile, more easily fitting with a human 'pack', were the ones more likely to stay around and more likely to be fed and encouraged. Over the years this selection became conscious and gradually the modern dog emerged.

What had been a natural process was transformed into genetic engineering, just as much as any GM crop. The dog is not a natural animal. It's as much a human-made piece of technology as a table that started off as a 'natural' piece of wood. Without doubt, the dog is one of the most impressive things our early ancestors made. Forget Stonehenge – it's a toy by comparison. Okay, it gave a handful of people some astronomical information, and it's pretty – but it hasn't been used for thousands of years. The dog is a piece of Stone Age technology, developed 35,000 years before Stonehenge, that is still going strong in airports around the world.

The security team might also use an electronic sniffer, which breaks down the chemicals in the air using one of a number of possible processes, most frequently gas chromatography. Here a gas carries the air up a tube, past various substances that the molecules in the air can interact with. Different molecules latch onto different substances within the tube. This splits out the components of the odour, so the machine can quickly produce a chart showing just what's in the substance it's sniffing. Different substances will have recognizable 'signatures' in the shape of their charts.

A lesson in detection

While your bags are being checked, you will have to pass through one of those intimidating arches that always make you feel nervous and guilty. These are metal detectors, similar technology to the hand-held devices used to hunt for treasure in a field, but here deployed to search *you* for metal. Although there are a number of variants, they all use the same basic process, called induction. If you have an electric toothbrush that charges by sitting on a plastic stand with no visible metal connectors, you already have a very obvious induction device in your home.

The idea for induction came out of a fundamental discovery made by the great Victorian scientist Michael Faraday. He discovered that moving a wire carrying electricity, or changing the rate at which the current flowed, produces magnetism. Similarly, moving or changing magnetism produces electricity. That's how electric motors and generators work.

In the toothbrush, a coil of wire in the charger sends out a changing electromagnetic field that produces a current in wires in the toothbrush. 'Electromagnetic' just means electrical and/or magnetic – electricity and magnetism are all part of the same phenomenon. And this 'field' is a field of force. This is a concept that Faraday dreamed up. He had seen how iron filings line up on a piece of paper held over a magnet, producing curved lines that seem to map out the magnet's invisible power. Faraday imagined these lines filling the space around a magnet.

Move a wire through a magnetic field and the wire hits line after line of the field, like a child's hand slapping a series of iron railings, transforming magnetism into electricity. There's no difference between moving a magnet next to a wire and moving a wire through a magnetic field – in both cases the result is a relative movement between the wire and the magnetic field, encouraging electrically-charged electrons to move in the wire. It's the same mechanism that appears in every generator.

In the toothbrush charger, nothing is moving, but the electrical current keeps changing direction (this is alternating current), making the lines of force shoot out from the charger and pull back in again. When a wire is positioned in the way of these moving lines of force it cuts through them, just as the moving wire does in a generator. There is no direct contact between the wires in the charger and the wires in the toothbrush. Instead it's magnetism generated by the changing electrical field that carries the power from the coil to generate electricity in the toothbrush. Similarly, devices called transformers that are used to drop voltage (you'll have several in your home in chargers and power supplies for mobile phones and electronic gadgets) do so by having a pair of coils of different size, where a changing current in one induces a current in the other via magnetic induction. (See page 126 for a definition of voltage.)

In the arch of a metal detector, there will be several coils of wire. The flow of electricity in these generates a magnetic field, which produces electric currents in any metal objects nearby. These currents generate magnetism in their turn, which finally produces electricity in a detection coil. The metal objects might be coins in your pocket, a belt buckle, or a weapon in your jacket. More recently, since shoes have been used to carry dangerous items, it's normal to take your shoes off to have them X-rayed, as the detectors can't cope with items at floor level, though some modern metal detectors can scan shoes, making the process less irritating.

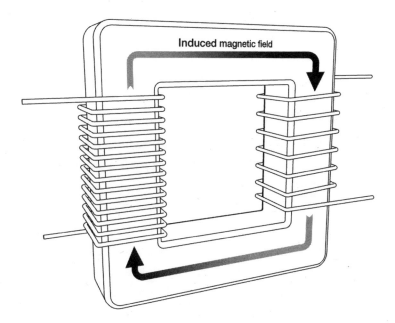

2. A transformer changes electrical voltage
via magnetic induction.

Body scan

There's an increasing possibility at airports of being subjected to a whole body scan. These scanners offer similar
facilities to a strip search, in that all kinds of items carried anywhere about the body can be detected, but the
process is quicker, taking only a few seconds, and feels
less intrusive. It has been said that such scanners produce a nude image of the individual, and as such violate
privacy, but in reality this is an exaggerated concern. The
result is unrecognizable – more like a computer model of
a human body than anything real.

There are two types of these scanners, both using
non-visible forms of light. Some employ high-energy

(short-wavelength) radio, while others use a form of X-ray. Small health concerns have been raised in some quarters about both of these. The radio version uses a frequency that is quite close to microwaves, and though there's no known evidence of a health risk from these radio signals, there are some concerns that the small levels of heating caused in human bodies might have a negative effect.

X-rays are known to carry a risk, but the approach in an X-ray scanner is totally different from a traditional medical X-ray. Full body scanners use a process called backscatter X-ray. Here, instead of passing through you, the X-rays pass through your clothes but are bounced back from your body to detectors all around you. These are very low-dose X-rays. You will be exposed to around 50 times as much damaging radiation for each hour you are in the air (we'll come back to this natural radiation later on) as you will receive from a backscatter scan. Overall the health risk is very low, and the process is significantly less unpleasant than an intimate body search.

Who do you think you are?

We now know that you are safe, but we might not be sure who you are. On international flights, after security, you will pass through border controls. Here, increasing use is being made of biometrics to identify individuals. Many passports contain a tiny chip that can be loaded up with your biometric data. These are simply measurements that can be checked on the day you travel to see if you really are the person your passport says you are.

Although in principle any aspect of your body could be used for biometrics (ear size, for instance), in practice most systems use one or more of face recognition, fingerprint recognition and iris recognition. Retinal scans, popular in spy movies, where an image is taken of the inside of your eyeball, have been avoided because the process feels too scary and intrusive. Few people, it seems, fancy having a laser blasted into their eye.

The idea of using fingerprints to recognize individuals is the most familiar technique, even if it does have unfortunate criminal associations. The first practical use of fingerprints seems to have been by Sir William Herschel, grandson of the astronomer of the same name. In the 1850s, he used them when working in India as a way to make clear identification on legal documents. By the 1890s they were starting to be used in criminal cases, with police departments building up libraries of prints, classified by shapes to make identification easier. From the beginning, though, matching a fingerprint found at a crime scene to an entry in a library was a tedious business.

Using fingerprints for biometric identification is much simpler as there's no need to search through a vast database. Instead it's just a matter of comparing the biometric data held on the passport with measurements taken on the day. Fingerprint identification technology picks up differences between the ridges and valleys in the skin of the fingertip, using a number of possible detection methods from a simple scan or the pattern of heat given off to electrical capacitance (the technology used on an iPhone touch screen). The whole print isn't stored – instead, key points in the pattern are identified, and compared with the stored data.

There are two problems with fingerprints. One is the need to capture a shape that can vary quite significantly over time, and that distorts depending on both the amount of pressure used and the position of the fingertip on the sensor. The other is the technique's criminal associations. It's very difficult to get someone to give a fingerprint without them feeling guilty. By contrast, iris recognition has none of those negative connotations.

The iris is the coloured bit around the pupil of your eye, which, on close examination, has a very detailed pattern of fine lines heading out from the centre like the spokes of a wheel. This unique pattern is captured by a camera and can provide an effective match to the data on the passport that isn't influenced by transparent materials like a pair of glasses. And there's no need to come into direct contact with the detection device.

Of the three options, face recognition is the ideal technology, because it can be undertaken remotely without the individual having to stop at a booth to give a fingerprint or an iris photograph. But as yet it isn't reliable enough to be the sole method of identification. Face recognition can be applied to a flow of people (though obviously it requires the face to be visible) or more practically for security, it can be performed unobtrusively in the background at any point an individual stops to speak to an official.

There are a number of ways that face recognition can work – picking up the locations of main facial features, taking a 3D scan of the face shape, or working more like fingerprint recognition on aspects of skin texture – but all are susceptible to variation, whether from an individual growing a beard or even from major changes in expression. The technology is a work in progress, but it provides a powerful extra check that is likely to become the dominant recognition mechanism as systems get better. Like it or not, your face says a lot about you.

The science of superstition

When the gate has been announced, and you've had enough of relative freedom, it's time to make your way to the holding pen that is the gate lounge. Gates are traditionally numbered, and you will usually find that gate 13 is missing. Although few people truly suffer from triskaidekaphobia – an irrational fear of the number 13 – the number is still often regarded as unlucky, something airlines and airports are enthusiastic to avoid.

The science of superstition is very much tied in with our perception of chance. Our brains just aren't wired well to cope with probability. You can see this with the way we react to clusters of events. Imagine something that happens randomly across the country, anything from outbreaks of disease to people falling over. How would you expect those random things to be distributed? Our natural response is to expect them to be spread out evenly. But that is totally wrong.

Just imagine tipping a tin of ball bearings onto a flat, empty floor. What would you think if, when the balls stopped moving, they were evenly distributed in a grid, each the same space from the other? You'd think something was making them do it – there must be magnets under the floor, or some other trickery. The natural thing is for there to be places where there are clumps of ball bearings and others where there are gaps. These clumps are known as clusters.

Experiment – Clustering cash

The chances are you don't have a tin of ball bearings about your person, and even if you did, you'd probably get arrested for scattering them across the floor of the plane. But you can get a similar effect with a handful of coins, though it's still best to wait until you get back home before you do this. Hold the coins in your hand about waist height and drop them. While in principle they could all fall nice and evenly, the chances are high that you will get clusters in their distribution.

Exactly the same thing happens with any randomly distributed event. But traditionally when, say, a number of farmers in the same area had cattle go ill, it was assumed that there had to be a cause for this clustering. The problems would be blamed on the local witch. Now, clusters of non-transmitted illnesses are often blamed on phone masts or nuclear power stations.

If such illnesses were random, we would expect them to form clusters; but it's very natural to look for a local cause, and, where there's a source of concern nearby, to assume that it's responsible. Not all clusters are random – a cluster of asbestosis victims near an asbestos factory, for example. But we can't assume that the apparent threat causes the problem. There are very effective statistical techniques to check on causality that need to be applied before jumping to conclusions.

Although people have put forward many explanations for the fear of 13, linking it to Judas as the 13th person at the Last Supper, or to the way 13 is the outrider number on collections of 12, like the 12 signs of the zodiac, there is very little evidence to support these theories. It's more likely that the number 13 was associated naturally with a cluster of bad events. Perhaps a farm failed after a sow had 13 piglets. Then, by coincidence, someone died on the 13th of the month. As a few coincidences built up, 13 would become the number everyone loved to hate.

Irrational though any fear of 13 is, airlines and airports don't take any chances of scaring their passengers, so there's unlikely to be a flight 13 or a gate 13. This avoidance is taken one step further at Heathrow's Terminal Four. Sometimes, when gate 13 is missed out, there's a tendency to consider gate 14 unlucky because 'it's really gate 13'. To prevent this from happening, gate 12 is at one end of the Terminal Four building while gate 14 is at the other end. As you never see the two gates side by side, it's not obvious that gate 13 isn't there, so no one worries about using gate 14.

Taking to the Sky

Aircraft basics

Whichever gate you were sent to, you will (hopefully) soon be directed to board and take your seat. It's a chance to have a look at the aircraft and notice a few details. We're so used to planes, we often don't realize what an impressive piece of technology a modern airliner is. Imagine you were with the Wright Brothers at Kitty Hawk in 1903. Their tiny spruce and muslin Wright Flyer weighed less than 300 kilograms (about the weight of a high-performance motorcycle) and had wings around 12 metres across. Compare this with a Boeing 747 weighing in at around 175 tonnes (and that's without any people or cargo on board) and with a wingspan of over 60 metres. That wingspan is nearly twice the distance of the Wright Flyer's 37-metre maiden flight.

All airliners have broadly the same configuration. They may be single- or double-decked, and have between two and four engines, but the design essentials are the same. A long, roughly cylindrical tube (the fuselage) in which passengers and freight are housed, which will have a rounded front end to reduce wind resistance. Part way down the body will be a pair of wings for lift (more on that later), with moveable segments for flight control. And down the back end there will be both horizontal (tailplane) and vertical (tailfin) surfaces sticking out to stabilize the plane in flight. These will have moveable sections for changing direction.

3. The undercarriage of an Airbus A-380.

Down below is the undercarriage or landing gear. These wheels are retracted during flight to make the shape of the fuselage more aerodynamic. Compared with an ordinary road vehicle there seems to be a mass of wheels on a plane. On a 747, for example, the landing gear has eighteen wheels in five units. But bear in mind that these tyres have to cope with a weight of up to 400 tonnes fully loaded, hitting the tarmac at around 150 miles per hour (240 kilometres per hour).

Fuelling flight

From the gate lounge or from your seat, you may see a plane taking on fuel from one of the fuel tankers, sometimes called bowsers. Aviation fuel, with that distinctive smell you are likely to have scented as you boarded, is a form of kerosene (paraffin). Like diesel or petrol (gasoline), it's a mix of hydrocarbons. These are organic molecules originating from crude oil that contain carbon and hydrogen, which have the useful trait of burning efficiently and producing a considerable amount of energy for their weight. Aircraft fuel typically has larger molecules than the petrol or diesel used for cars, and because of this is less volatile.

To get a feel for the kind of chemical we're dealing with, meet octane. Like all molecules, octane consists of a number of atoms stuck together by interactions of the charged particles that make the atoms up. If you could see a molecule of octane, it forms a long string of eight carbon atoms with a total of eighteen hydrogen atoms attached. We're familiar with the name 'octane' from the rating on petrol that gives us the expression 'high octane'. This has nothing to do with the amount of octane in the fuel, but is instead a measure of the fuel's anti-knocking capabilities compared to a standard formulation of fuel that contains octane. (Knocking is a kind of misfire when some of the fuel ignites at the wrong point in the engine's cycle.)

Fuel oil has a huge benefit for the airline industry. Weight is crucial for operating a plane, and aviation fuel packs in an immense amount of energy per unit of weight. Just look at the difference between fuel oil and batteries. Assuming we're using hi-tech computer batteries, to carry the same amount of energy as 10 kilograms of aviation fuel you need around a tonne of batteries. This is why you aren't going to see electric airliners any time soon.

Aviation fuel crams the energy in so well that it has fifteen times the energy per kilogram of the explosive

TNT. That might seem crazy, but the reason TNT is an explosive is not so much the amount of energy that is stored in it, as the speed with which it burns. Although a stick of TNT releases a lot less energy than the same weight of aviation fuel, it does so in a tiny fraction of a second. When TNT ignites it produces a huge burst of heat, generating a pressure wave in the air, and it's this pressure that causes the explosive damage.

Unlike road transport, or electric power generation, it's going to be very difficult to move aircraft away from fossil fuels to a cleaner energy source. One possibility is hydrogen. The simplest of the elements, hydrogen isn't an energy source in its own right, because you have to make it before you can use it. But it's an alternative way of transporting energy, which has the advantage over fuel oil that when it burns, the only emission is water vapour.

With enough electric power, you need only water to make hydrogen. As long as that electricity is generated from a clean source, the hydrogen is a green fuel. The great thing about this simple gas is that it packs in even more energy per kilogram than aviation fuel – nearly three times as much. But there is a problem. Hydrogen is bulky. It may be lighter than fuel oil, but as a compressed gas it takes up six times as much room as the traditional fuel, and tucking that away on an already cramped aircraft wouldn't be easy.

It may be that as oil becomes scarcer it's increasingly conserved for flying, though in the last resort there's a technique called the Fischer-Tropsch process that can convert coal into fuel oil. It was developed by the Germans during the Second World War, when they were cut off from their usual sources of oil. This process is important because the USA, for example, has several hundred years' worth of coal reserves it could call on if oil became difficult to obtain.

The reason the process isn't used so far is partly because it's dirty, and would need considerable development to reduce carbon emissions, and partly because the plants are very expensive to build – though once constructed, oil can be produced at around $50 a barrel, considerably less than the average price between 2005 and 2011.

There has been much discussion about how bad carbon emissions from flying are. The reason is that carbon dioxide is a greenhouse gas, which contributes to global warming. But in itself, carbon is not a bad thing. In fact this straightforward element is an essential for life. It's very good at combining with other elements to make long chains of molecules – and without that it would be impossible to build the proteins, DNA and other complex molecules that make life possible. Without carbon, we wouldn't be here.

This doesn't take away the fact that carbon dioxide is a greenhouse gas, though. And the greenhouse effect has to be bad, doesn't it?

The greenhouse effect's good side

We're used to being told how terrible the greenhouse effect is. We hear practically every day that we need to cut down our carbon dioxide emissions, because CO_2 is a greenhouse gas, and this is causing global warming. The warnings are real – there is too much carbon dioxide in

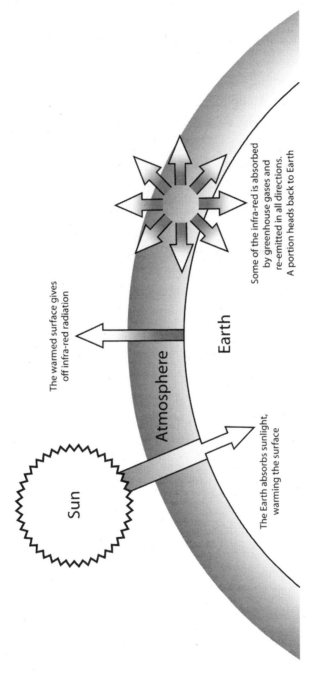

The warmed surface gives off infra-red radiation

Atmosphere

Earth

Some of the infra-red is absorbed by greenhouse gases and re-emitted in all directions. A portion heads back to Earth

Sun

The Earth absorbs sunlight, warming the surface

4. The greenhouse effect: greenhouse gases act like a one-way mirror.

the atmosphere at the moment – but we shouldn't think that the greenhouse effect is intrinsically a bad thing. It helps keep us alive.

In the greenhouse effect, carbon dioxide, alongside other greenhouse gases like water vapour and methane, acts like a one-way mirror. Most incoming sunlight shoots straight through it, but when the energy is re-emitted by the Earth as infra-red, which is lower-energy than the visible light that comes in from the Sun, some of it is absorbed by the CO_2 molecules in the atmosphere. Almost immediately the molecules release the energy again. A portion continues into space, but the rest returns to Earth, warming the surface.

An example of an out-of-control greenhouse effect is the planet Venus, with a 97 per cent carbon dioxide atmosphere. Here average temperatures are 480° Celsius, with peaks of 600°C, making it the hottest planet in the solar system.

Before we started pumping out CO_2 into the atmosphere, our greenhouse effect was operating at just about the right level. Without it, the average temperature on the Earth would be −18°C, over 30°C lower than it actually is. This is so cold that life as we know it would never have come into existence. Without the greenhouse effect, about the only place on Earth where there probably would be life is around the hot vents at the bottom of the oceans.

Flying the green way

There's no escaping the fact that flying is the worst means of transport for carbon dioxide emissions – and good though the greenhouse effect is, we don't want any more greenhouse gases. A typical long-haul flight from Europe to the US will produce around 2.5 tonnes of CO_2 per passenger. That's the equivalent of driving around 15,000 kilometres in an average car. If you're flying business class it's more like 4 tonnes, and for the first-class passenger, 5 tonnes. (That's because these seats take up a higher percentage of the space on the aircraft than those of ordinary travellers.)

If you want to do something to counteract the effects of air travel, look into carbon offsetting schemes that involve building renewable energy sources like wind turbines or wave power generators, preferably somewhere where the current energy sources are particularly dirty. This is significantly better than tree-planting. Tree planting is undoubtedly worth doing in terms of biodiversity, but in practice trees absorb carbon very slowly – we need the carbon reduction now, rather than in 100 years' time – and sadly trees often die, at which point they start to give off carbon rather than take it out of the atmosphere.

Getting moving

After what can seem a long wait, it's time for pushback. The aircraft reverses away from the stand and taxies off to the end of the runway. Unlike a car, there's no power to the wheels in a plane; most of the manoeuvring on the ground is driven by the aircraft engines. This is not a very efficient way to travel when the plane isn't in the air, particularly in reverse, so to get away from the terminal an aircraft tug (sometimes called a pushback tractor) is usually brought in.

5. A tug or pushback tractor in action.

These squat, low-profile vehicles, which are also employed to manoeuvre aircraft when they are not in use, aren't as powerful as you might imagine. The tugs used on 747s are typically 200–300 horsepower – less than a high-performance car. We tend to come across this unit of horsepower only when we're looking at means of transport. It's a measure of power that was devised by Scottish engineer James Watt.

Words like 'power' and 'energy' are used pretty interchangeably. If we say someone has lots of energy, it means they have lots of ability to get things done; while if we say someone is powerful, they have the capability to make things happen. In science, though, the terms have very precise meanings. Energy is the ability to do work, it's the oomph that makes things happen. Power is the rate at which you do work (or provide energy) – it's the amount of energy used per second.

We usually measure power in watts, named after James – so you might have a 100-watt bulb, which uses up 100 units of energy (joules) a second, or a 2-kilowatt (2,000 watts) kettle – but horsepower is just an alternative unit. When Watt devised the horsepower, he was trying to provide a way to compare the rate at which a steam engine worked against the rate a horse worked. He measured how much work a horse typically put in during a shift in a mine, then arbitrarily doubled it to produce the horsepower. One horsepower is around 750 watts, or three quarters of a kilowatt.

So aircraft tugs aren't ridiculously powerful (a large truck can easily have twice the horsepower), but they are immensely heavy, weighing as much as 50 tonnes. Having this great weight pushing down on the tyres means that the tugs get a huge amount of traction, and they are geared for plenty of torque, the turning force that powers the wheels, when moving at low speed. The result is that they find it easy to get several hundred tonnes of aircraft moving.

In principle, an airliner can back away from the terminal using reverse thrust. This involves the crude technique of placing a deflector behind the jet engines, so the blast of air is pushed towards the front of the plane. Reverse thrust is usually deployed on landing to slow the aircraft down – this is what is being engaged when you hear the engines suddenly surge as you touch the ground. But it isn't practical to use reverse thrust when close to a terminal ('on stand' in airline parlance). The blast from the engines is liable to send any debris on the ground hurtling towards the glass of the building, which is why tugs are used instead.

You may wonder, given the inefficiency of taxiing on jet engines, why the tug doesn't take the plane all the way to the runway. Virgin Atlantic did come up with the idea of doing just this in 2006. The idea was to pull the plane to a 'starting grid' at the end of the runway. This would have produced significant fuel savings – Virgin reckoned that they could save two tonnes of CO_2 per flight, as well as reducing noise and cleaning up the air near the terminal.

Unfortunately, despite its green credentials, the technique soon had to be shelved. This was partly because airports were not willing to provide the starting grid locations, which would have produced delays while tugs were decoupled and moved clear of the jet blast. But more significantly, the aircraft manufacturers warned that increasing the amount of towing would put too much strain on the undercarriage, meaning the struts that hold the wheels would have to be replaced more frequently. The chances are, then, that you will taxi to the runway using the plane's engines.

Big radar is watching you

As you traverse the sometimes labyrinthine complex of taxiways at larger airports you will be tracked by surface movement or ground radar, just the first of many encounters with radar that your aircraft will have along its journey, as it's handed from ground control to the tower to terminal control and finally on to area control. The last two of these are the air traffic control centres that will monitor your flight once it's in the air. At most airports you will see at least one radar aerial sweeping around. And your plane itself will have a radar system, built into the nose, that it uses to warn of storms ahead.

Although there were various attempts at producing systems to detect aircraft earlier, the first practical technology for spotting planes beyond the limits of visibility was developed in Britain just before the Second World War. Originally called 'range and direction finding', it was soon renamed radar, the less cumbersome American acronym for Radio Detection and Ranging. According to radar legend, the technology was devised when British experts from the Radio Research Station at Slough were asked to look into claims that US/Croatian inventor Nikola Tesla had devised a death ray based on electromagnetic radiation. The scientists found it highly unlikely that radio could be used as an alternative to bullets, but they did think that the right sort of radio beams could be effective for detecting enemy aircraft.

The early British work with radar seems to have been the source of the old story that eating carrots improves your eyesight, especially in the dark. This was a piece of surprisingly successful propaganda put out by the Air Ministry during the war, who claimed that fighter pilots were on a diet that was rich in carrots to improve their night-time eyesight so that they could spot incoming German bombers. In reality it was radar that was helping direct the fighters towards their targets, but it was hoped that the Germans would believe the stories. There was so much coverage in the British press that it became common folklore that carrots help you see in the dark.

The principle of radar is very simple. It works using a form of light sitting on the electromagnetic spectrum between the frequency used for radio and TV and that used in microwaves. As we discovered earlier, when we see an object using visible light, the photons of light travel from the source – the Sun, for example – to the object. The object absorbs the photons, pushing up the energy of electrons in the object's surface. Soon those electrons drop back down in energy and new photons of light are emitted. Some of these reach our eyes and we see the object.

Radar works quite similarly, except that it's both the source – the equivalent of the Sun – and the detector – like our eyes. It emits a stream of photons, then looks out for the re-emitted photons coming back from an object. Because the photons have lower energy than visible light (if we think of light as a wave, the radar waves have longer wavelengths, making bigger ripples as they travel along), they are less able to detect specific features, so typically just show a blob, rather than the detailed view we get with eyesight.

Something on the air tonight

Radar isn't, of course, the only way that electromagnetic radiation will help your pilots get you to your destination. Apart from using visible light to see where they are going, they will make wide use of radio, a form of light with even lower energy per photon (longer wavelengths) than radar. Some of the radio use is automated. Navigation beacons are unmanned radio transmitters that send out a continuous stream of information to enable planes to know where to turn on flight paths. Although their use has become less significant with the introduction of GPS (see below), they are still valuable guides. Called VOR (VHF Omnidirectional Radio Range), a plane typically uses two of these beacons to get a fix on location.

A more sophisticated collection of automated radio transmitters at airports provide ILS (Instrument Landing System) facilities, which pinpoint the line of the runway and the angle at which the plane is approaching the ground. By combining ILS information with special radar facilities, a suitably equipped plane can land in zero visibility without the pilot handling the controls, a system known as autoland.

Before sophisticated radio-based navigation and landing systems, pilots had to rely on visual indicators to guide them into an airport. To line up ready for the runways they would use landmarks that could be easily recognized from the air. One of the standard markers, for example, on the route from the east into London's Heathrow was a large gasholder. Unfortunately there was a similar gasholder with the same orientation to the runway of the nearby RAF base at Northolt. The pilot of a 707 airliner belonging to a US airline once confused the two gasholders and touched down at Northolt, believing it to be Heathrow. This caused a serious problem. An aircraft needs a longer runway to take off than it does to land. The Northolt runway was too short for a 707 to get into the air.

After stripping everything from seats to galleys out of the plane they just managed to take off. Local mythology says that there were tyre marks on the roof of an office block situated near the end of the Northolt runway, left as the airliner clawed its way into the sky. After this, both gasholders were given markings to identify them. The Heathrow gasholder, for example, was painted with the letters LH – you can still see it from the train as you pass through Southall in west London.

Experiment – Landing on Google

If you have access to the internet at the moment, you can recreate the experience of using the correct gasholder to identify your flight path by taking a look on Google. Go to http://maps.google.co.uk and search for Southbridge Way, Southall. Drop the 'Street View' man (the little orange man at the top of the scale gauge) on the left-hand end of Southbridge Way. When you get a picture of the area, turn so you are facing roughly west; you will see the light blue gasholder with LH painted on the top to indicate that this is the Heathrow gasholder, not the Northolt one.

Sat nav on the flight deck

Now, though, just as with a car, modern navigation in the air is mostly in the hands of another radio-based system: GPS (Global Positioning System) or sat nav. This uses the nearest handful from a collection of between 24 and 30 satellites scattered around the globe at an altitude of 20,000 kilometres (12,500 miles) to pinpoint the location of a GPS receiver anywhere on the Earth. (More satellites are being added to make the system more accurate, so the number is increasing.) Each satellite carries a very accurate clock and constantly broadcasts the time and the satellite's orbital position. The GPS receiver typically latches onto between four and six of these transmissions and uses them to fix its location based on knowledge of how long the signals from the satellite should take to arrive at the speed of light.

GPS provides a living demonstration of the realities of Einstein's theory of relativity. Einstein came up with two forms of relativity: special relativity, which describes how time and space are influenced by movement; and general relativity, which looks at the impact of acceleration and gravity. We'll revisit what relativity tells us in detail later, but GPS is a practical example of the need to take relativity into account. Relativity isn't just an arcane theory, it has a direct influence on this everyday piece of technology.

Special relativity says that fast-moving clocks will run slower than you would expect them to – and predicts that the clocks on GPS satellites will lose around 7 millionths of a second every day compared with clocks on Earth because they are travelling at 8,600 miles per hour. General relativity tells us that gravity also has the effect of slowing clocks down. As the satellites experience a weaker gravitational pull than we do on the Earth's surface, their clocks should *gain* around 45 millionths of a second a day. Overall, the result is clocks on GPS satellites gaining around 38 millionths of a second – and they do. This amount of error sounds trivial, but satellite navigation is dependent on very accurate measurements to get the position right. Without correcting for relativity, the GPS system would soon become useless. In just a day, without correcting for relativity, the position provided by GPS would be wrong by several kilometres.

The universal language

It's radio that provides the link to the GPS satellites, and radio that enables planes to talk to each other and to communicate with controllers on the ground. It was realized early on in the history of flight that it could be dangerous if instructions from a controller to a flight couldn't be understood by a second plane from another country. For this reason, all commercial air traffic communication is undertaken in English – even if it involves, for example, a Chinese controller handling a Chinese flight.

Unlike light aircraft, which are indentified by the registration number of the plane, a commercial airliner will be referred to on the radio by a combination of a designator for the airline plus the numerical part of the flight number. Sometimes these designators are obvious and predictable. American Airlines, for instance is AMERICAN and Qantas is QANTAS. But others have more obscure designators. British Airways, for example, is SPEEDBIRD (after the logo first used by Imperial Airways and still displayed on the side of the aircraft), while a small British operator called Special Scope has the less than flattering designator DOPE. There are even flights called SANTA – Christmas charter excursions operated by BA.

The latest model on the runway

By now you should have reached the beginning of the runway. At a major international airport, the runways can be 3 to 5 kilometres (1.8 to 3 miles) long. Big airports often have parallel runways to increase throughput, with additional secondary runways positioned at different angles. This is because aircraft take-offs and landings are best made facing into the wind.

It might seem that the last thing you want is the wind pushing against you when you're trying to get up to speed, but there is a useful benefit. If you take off into the wind, for any particular speed over the land, the air is moving faster over the wings – it flows at the combination of the plane's speed and that of the wind.

Say a plane needed to be travelling at 150 miles per hour to take off. With a 50mph headwind, it would need to travel at only 100mph over the ground – but with a 50mph wind behind it, reducing the speed of the air over the wings, the plane would have to reach 200mph along the ground to get into the air.

Realistically, you can't have runways facing in all directions, so airports typically go for the prevailing wind direction. The runways are labelled (you'll see a large number painted on the end of the runway) with a contraction of their compass direction. If the direction of the runway is within the first ten degrees to the east of north, it's designated 01. The next ten degrees is 02 and so on. As planes may have to approach the runway from either direction, depending on the wind, the two ends of the runway will be labelled with numbers that differ by 18 (because they're 180 degrees apart).

For instance, London Heathrow has two parallel east–west runways designated 27 Left and 27 Right, or 09 Right and 09 Left, depending on the direction of approach. They're 27 if you're heading west from the London direction, and 09 if you're heading east. The airport did have a shorter third runway crossing these at an angle – known as 23 or 05 – but this wasn't long enough for large jets and was closed in 2005. (You can still see it in online aerial photographs like Google Maps; it is used as a taxiway. The numbers at the ends of the runway are painted out but are clearly visible.) Other airports have considerably more runways – Chicago's massive O'Hare airport, for example, has seven.

When you reach the start of the runway, there are two ways that the pilot can act. Their preference is a rolling start. Here, the plane turns onto the runway and immediately, without stopping, the throttles are opened and the plane begins its take-off run. This is more efficient, as the engines don't have to get the plane going from a standstill, and it also has a decidedly macho appeal that wins over most pilots. But there's a fair chance that you will have to wait in place for a number of minutes.

This delay is most likely to occur when there's a string of flights taking off. It's not just a matter of queuing until the plane ahead has started down the runway, so you don't get hit by its jet blast – the wait is significantly longer, particularly if the aircraft in front is bigger than yours. The reason is that the wingtips of the aircraft in front can generate powerful spinning vortices in the air. Think of the way that water spirals down a plughole in a sink – this is a tiny vortex, though the vortex from the wingtips is invisible and much more powerful.

Experiment – Within the vortex

To see a vortex in action at home, put the plug in a sink and fill it up as far as you can. Carefully draw the plug out and watch the water leave (it's better in a bath, as there's longer for a vortex to form). You should see a little whirlpool on the surface of the water, heading towards the plughole – this is the vortex.

You may have heard that these plughole vortices go in different directions depending on which side of the equator you are. This is supposed to be because of the Coriolis force, a real effect caused by the spinning of the Earth. If you think of someone standing on the North Pole, then they effectively don't move forwards as a result of the spin of the Earth, they go round in a circle. Head towards the equator and you move forwards with the Earth faster and faster, because you are heading around a bigger circle at the same rotational speed. So if you have something like the water in a bath, the part of it nearer the pole is moving slower – the effect, if the object is not fixed to the Earth's surface, will be to rotate it in a clockwise direction.

This Coriolis effect can be seen on weather patterns, which often rotate clockwise in the Northern hemisphere and anti-clockwise in the South. But the amount of force across the size of a sink or bath is so tiny that it has no influence on the direction of the vortex. Instead, this is determined by the way you pull out the plug and other physical factors like the shape of the edge of the plughole. These far outweigh the Coriolis effect.

The vortices produced by the wingtips of a plane can take two to three minutes to subside. If another plane flies into the disrupted air it can become difficult to handle, so the gap between take-offs allows enough time for the air to settle down.

How Newton's laws get you going

Whether you have had to wait or turned straight onto the runway, the moment will come when the throttles are opened and you are thrust back into your seat. Here's a chance to experience one of the most famous bits of science in action – you're in the hands of Newton's laws of motion. These laws are all about forces, and in essence, a force is just something that makes something happen. If an object suddenly starts moving, for instance, a force has to have been applied to it. Newton described how all this works in three separate laws, all of which come into play as you start rolling down the runway.

The first law says that an object (your body, for example) will stay the way it is – moving or still – until a force is applied to it. This sounds trivial, but before Newton, the assumption was that you had to keep pushing something to keep it moving. If you stopped pushing, they thought, it naturally stopped. (In fact, the old idea was a bit more complicated, as they thought that some things (like earth) had gravity, which was a natural tendency to fall towards the centre of the universe, and other things (like air) had levity, which was a natural tendency to rise away from the centre of the universe, but otherwise things stopped moving if you weren't pushing them.) Newton, however, realized that once something is moving, you need to apply a force to it (in the opposite direction to its movement) to slow it down.

The jet engines apply a force to the plane, and that force gets the aircraft in motion. But your body isn't moving at this point. So the seat pushes forwards on you. What you feel (because you see things from your point of view, rather than the viewpoint of the seat) is that you are sinking back into your seat, but in terms of cause and effect it's your seat that is being pushed forward into you. A force is applied to you – and you start moving. It's just as well, as if you didn't start to move you would crash through the seat as it travels forward.

So what does this force do to you exactly? Newton's second law tells us that the amount of force that is being applied to you is your mass times the acceleration you experience. The more force you feel, the faster you will accelerate. The plane will accelerate from a standstill to around 150 miles per hour in something like half a minute. That is around one quarter of the acceleration we experience due to gravity. So the g-force you experience – the equivalent of the force of gravity – is around 0.25g.

This seems pretty feeble. And compared with the 0 to 60 acceleration of a performance car, it is. In a Jaguar XJR, for instance, you should be able to get from 0 to 60mph in five seconds. That's around 0.6g. So why is it that you are so noticeably pushed back into your seat on the plane? The moment the plane's engines throttle up, you're hit by the majority of the thrust, while the force builds more gradually in the car, so it isn't always as noticeable (and most of us don't drive high-performance cars).

The third and final of Newton's laws plays a big part in your acceleration down the runway. This is the one that's often phrased as 'every action has an equal and opposite reaction'. On the face of it, that sounds like rubbish. It sounds, for instance, as if you could never move anything, because when you try to push it, there's an equal reaction pushing the other way – net result, nothing should happen. Yet without Newton's third law, a jet aircraft couldn't move.

The reason the third law works is that the action and the reaction apply to different things. When you push a box, the box pushes you back with an equal amount of force. If you go sky-diving, the Earth is attracted to you with exactly the same force as you are attracted to the Earth. But remember the second law. Force is mass times acceleration. The Earth has a much bigger mass than you do. So though both you and the planet experience the same force due to gravity, the acceleration the Earth feels is that force divided by a huge mass – to all intents and purposes it doesn't move.

When your seat pushes you, you push your seat back. If you didn't, you wouldn't sink into the seat with the acceleration, you would just be shoved forward. But Newton's third law is much more important than this. A jet engine depends on that law to move the plane.

Joining the jet set

The big fan you can see at the front of a stopped engine sucks in air and compresses it. The air then gets mixed up with a mist of fuel and the combination is ignited in a combustion chamber. This produces a blast of energy, part of which powers a turbine to keep the compressor blades going, but most of which shoots out of the rear of the engine, joining the powerful flow of air sucked through by the compressor blades. Because the engine gives the air a huge push backwards, in its turn that air gives the engine (and hence the plane) the same amount of thrust forwards. The only thing that keeps the plane moving is Newton's third law.

Jets generate a huge amount of thrust. The combined force produced by all four engines of a 747 can be as much as 1,000 kilonewtons, or 1 million newtons. A newton (yes, named after him) is the unit of force. It's the amount of force needed to accelerate one kilogram by one metre per second in a second. Compare this with an extremely high-performance car like the fastest Porsche, which can accelerate from 0 to 60 miles per hour in three seconds. If the car's weight (including the effect of friction) is two tonnes, that would take just 18 kilonewtons of thrust.

Most of the time, Newton's laws do the job when we're working out how things move, but technically the second law is an approximation that works only when something is moving relatively slowly compared to the speed of light. The faster you go, the less accurate it is, and even at slower speeds it isn't quite right, which is why GPS satellite timings need to be corrected for the inaccuracy. As we will see later, for absolute accuracy – or for really fast things – we need to replace Newton's second law with the predictions of relativity. But for the vast majority of everyday uses, Newton does just fine.

Rotation and climbing

Up front, as the plane accelerates, the pilot will be told as the plane passes three critical air speeds, known as V1, VR and V2. Each of these speeds is fixed for a particular aircraft to maximize safety during takeoff. V1 is, in effect, the speed of no return. When the first officer calls out 'V1' the plane is committed to take-off. The 'R' in VR stands for 'rotation', and at this point the first officer calls out 'Rotate'. Now the pilot pulls back gently on the control stick, changing the angle of the tailplane control surface so that the aircraft rotates on its main under-carriage, lifting the nose wheel off the ground. This tilt means that the air is hitting the wings at a sharper angle, giving more lift (see below). The plane is still accelerating on the ground until V2 is reached, the take-off speed.

First-time flyers are often unnerved by the way shortly after take-off the engine noise usually drops away. There's nothing wrong; this is standard practice. There's still plenty of power being applied to continue climbing and accelerating, but by pulling back a little on the throttle, the amount of engine noise is cut, reducing the nuisance for people who live near the airport.

Under pressure

As the aircraft climbs, you will usually get some sensation in your ears, which could range from a mild feeling of pressure to a sharp pain. This ear popping is caused by the air pressure in the cabin of the aircraft reducing to a level below standard atmospheric pressure. A commercial airliner cruises at between 35,000 and 40,000 feet (around 7 miles or 11 kilometres). This is partly to save on fuel because of the reduced air resistance, and partly to be well above most of the weather systems that can make a flight uncomfortable. At that level, the air pressure outside is too low to be breathable, with around a quarter as much oxygen as you'd get on the ground, so the cabin is pressurized.

In principle the cabin could have the same pressure as you would have at sea level, but the greater the pressure, the heavier the plane has to be to stay airtight – there's a trade-off. It was arbitrarily decided that a reasonable cabin pressure was the equivalent of being at between 6,000 and 8,000 feet above sea level, reflecting the altitude of the highest cities on Earth like Mexico City. The pressure at 6,000 feet is around 80 per cent that at sea level, meaning you get around four fifths as much oxygen as usual. At 8,000 feet you're down to around 75 per cent normal pressure.

This reduced pressure may make you feel a trifle breathless and easily tired, but the main impact that the cabin pressure has is on your ears. As the air pressure around you decreases, any pockets of gas inside your body will expand. The discomfort you may well feel is caused by the air in the twisting Eustachian tube that links your nose to your mouth. As the pressure outside drops, the air pocket inside expands, pushing against your ear drum, causing discomfort until you can equalize the pressure by swallowing, yawning or the Valsalva manoeuvre, which involves holding the nose and gently blowing.

Wing work

The jet engines alone aren't enough to get you off the ground – it's the wings you have to thank. Just moving forward at high speed won't stop you from falling back to earth. Imagine for a moment that you're on the ground with a gun in your hand, while holding in the other hand an identical bullet to the one that's in the gun. You simultaneously drop the bullet from your hand and fire the gun horizontally. Which of the two bullets hits the ground first? The natural inclination is to say that it will be the bullet in your hand, but in fact both bullets will hit the ground at the same time. The bullet shot from the gun falls at exactly the same speed as the stationary bullet.

So speed alone isn't enough. You need lift – an upwards force to counter gravity and lift the plane off the ground. This is the job of the wings. When a bird's wings are flapping, the source of that upward thrust is fairly obvious. The flapping wing pushes air downwards, and just like the jet engine, the result is a Newton's third law upward push on the wing, lifting the bird. But what happens when a bird is gliding, or to lift up a plane, where the wings are fixed in place and don't flap to push at the air? The effect is initially quite counter-intuitive.

Experiment – Give yourself a lift

It's easy to recreate the effect that lifts your plane into the sky. Tear off a strip of paper. If you take a sheet of A4 or Letter paper and fold it in half along its longest dimension, then fold it in half again in the same direction, you will produce an ideal sheet. Tear it off and hold it by one end so the paper droops away from you. Now put the end you're holding up to your face immediately below your lips. Finally, blow a long, steady breath across the top of the paper.

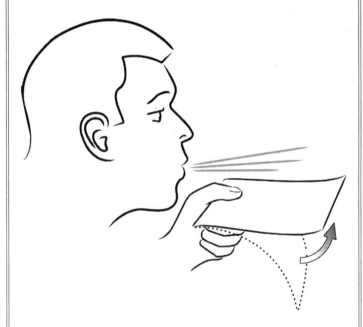

6. Producing lift in a sheet of paper.

Your paper strip will lift up, no longer drooping. You have produced lift in it, just as an aircraft wing produces lift. This lift, as we will see, is caused by the air moving differently over the top of the lifting surface – in your case the paper – to the way it moves over the bottom. In your paper model, the air isn't moving over the bottom at all, but it's rather different for a plane's wing.

The wing is shaped with a drooping curved surface on top. Just like your sheet of paper, when air blows over this curved shape, the result is to produce a lifting force, pushing the wing upwards. What is happening is that the shape of the wing, called an aerofoil or airfoil, cuts through the atmosphere and turns the flow of air in a different direction. As the aerofoil applies a force to the air (it's Newton's third law again) the wing is pushed in the opposite direction. We'll see in a moment why this happens, but embarrassingly, for a long time the most common explanation of how aircraft lift works was wrong.

You may have heard that wings have a special shape whereby the length of the journey that the air travels over the top of the wing is greater than the distance along the bottom. As the air has further to go along the top, the argument goes, these air molecules will travel faster to keep up with the air going around the bottom, which thins out the air across the top of the wing. And with less air at any particular point, the pressure on the top surface drops. That means the wing should feel a force upwards.

It's true that if the air does move faster over the top of the wing you will get lift, but the difference in length of path on the two possible routes around the wing has nothing to do with it. There's no particular reason why air going over the top would try to catch up with air going around the bottom. As it happens, the speed that air goes over the top of a typical wing shape is much faster than the speed necessary to catch up with air going around the bottom. This effect is completely unrelated to the different distances along the top and bottom of the aerofoil. Instead it's down to the complex way a fluid like air moves.

To understand what's really happening we need to briefly revisit Newton's second law. This told us that force equals mass times acceleration. If there's acceleration, there has to be a force. But what is acceleration? We're used to thinking of it as a change in speed. Going from 0 to 60 in six seconds, for instance. But acceleration is actually a change in *velocity*, not a change in speed. The difference between speed and velocity is that velocity has both speed *and* direction. It's what's known as a vector. Any change in velocity is an acceleration, even if the speed stays the same and only the direction has altered. When something is going around in a circle at a constant speed, it is accelerating, and that acceleration needs a force to make it happen.

So imagine the air flowing around the wing. It changes direction because it's passing over the curved upper surface. This means the air is accelerating, and as it curves down over the wing, that acceleration is downwards. A force is being applied downwards on the air by the wing, and the air exerts an equal and opposite force up on the wing.

Control surfaces in action

A related effect is used to steer the plane using 'control surfaces'. These are moveable bits of the wing and tail section that can be used to tilt the aircraft and change direction. Each of the three main control surfaces acts the same way – the ailerons, which are long strips along the outer trailing edge of the wings, the rudder on the vertical tailfin, and the elevators on the horizontal tail-plane. Ailerons are used to bank the plane to left and right, providing the main turning mechanism in the air. They go in opposite directions when banking, one up and one down, with the rudder providing secondary turning control. The elevators both go the same way, tilting the aircraft up or down in the air. In each case, moving the control surface changes the forces operating on the plane in a similar fashion to the wing producing lift.

If you're seated in sight of the wings, you may also see extensions on the inner trailing edge of the wing retract soon after take-off – and you will certainly see them extend before landing. These are flaps. Flaps have a dual effect on the wing. By increasing the area of the wing, they push up the amount of upward thrust. This means that the plane can fly stably at a lower speed – essential for landings and often useful in take-off. But the flaps also increase drag. This is a backwards force that slows the plane down. Drag is useful when landing as a way of reducing the airspeed, but it's something you want to reduce in normal flight, so if the flaps were used in take-off they will soon be withdrawn to minimize drag. If you look under the wing you will usually see a series of elongated streamlined struts (called flap track fairings)

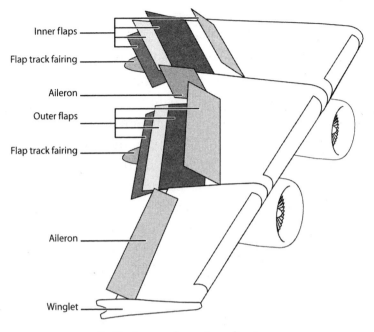

Inner flaps

Flap track fairing

Aileron

Outer flaps

Flap track fairing

Aileron

Winglet

7. Features of an aircraft wing.

– these house the mechanisms that extend and retract the flaps.

Another thing you will notice if you're in sight of the wings is that they aren't totally rigid. They flex as the plane travels. This isn't flapping like a bird's wings – the plane isn't generating lift this way – they are just flexing because it takes time for the upward and downward movements of the plane's body to move along the wings. The tips of a large airliner's wing can move 2 to 3 metres in flight, which looks extreme, but is relatively minor compared with their tolerance for bending. When the Boeing 787 was undergoing tests before flight, the wings were flexed by a remarkable 8 metres without failure. If a wing were totally rigid, it would put intolerable stresses on the junction between the wing and the fuselage – remember a fully loaded 747 can weigh over 400 tonnes. Flexibility enables the wing to take the strain.

On more modern planes you may also notice winglets – apparently small extensions to the wing sticking up from the tips. The small size is only relative to the huge wing – they are often as high above the wing surface as a person. The winglet has a double effect. It increases the amount of lift for a given wingspan, and it reduces drag. This is because the tips of the wings tend to produce turbulence in the form of disruptive vortices of air – the same whirling structures that mean a gap has to be left between planes taking off. Putting winglets on the wingtips slices through these vortices and reduces the effect, cutting down on drag. Some modern planes don't have obvious winglets, but instead have an increased slope on the outer section of the wing, which has a similar effect.

Experiment – Forces in flight

When a plane is in the air, it's subject to five main forces, providing a complex mix of inducements to move. These are:

- **Gravity** – pulling the whole plane downwards
- **Lift** – pushing up on the wings
- **Thrust** – pushing the plane forwards in response to the engines
- **Drag** – due to air resistance, pulling the plane back against the thrust and the lift

- **Turbulence** – pushing in all sorts of directions as the plane is exposed to various air currents

If you have three sheets of paper around A4/Letter size, you can demonstrate several of these effects. Your aim is to throw each piece of paper as far as you can (for reasons of sociability, wait until you're back on the ground).

Throw the first sheet unfolded. Throw the second scrumpled up into a ball. And throw the third after you have folded it into a paper airplane. Each should travel further than the previous one (unless you're bad at making paper planes). The gravitational force on the paper and the thrust (from your throw) should be roughly the same in each case. Unless your paper passes an air stream, there should be relatively little turbulence. So what changes between the three sheets is lift and drag.

The first, open sheet has maximum drag because it exposes a lot of surface area to the air. Drag is caused by the air molecules bashing against the sheet, producing a Newton's laws force. If there's a lot of surface area, more air molecules hit the sheet, so there's more drag. The second, scrumpled up sheet has less drag, because you have reduced the exposed surface area that will be hit by air molecules.

Both open sheet and scrumpled ball will have a little lift, but a properly constructed paper plane (which may have more drag than the crumpled ball) should have a lot more lift, enabling it to fly further than either of the other sheets.

Exploring the Landscape

The mystery of the fields

Once you've cleared the edge of the airport, you're in one of the most fascinating parts of the flight. Still low enough to see plenty of detail on the ground, you have a rare opportunity to take a bird's-eye view of your surroundings. These days, anyone can take a look at a static aerial view on the internet, but there's a realism and expansive scope of vision that you have from your aircraft window that enable you to take in the details of town and country afresh.

If you fly over fields in summer and early autumn, you may have a chance to see crop circles as they are best viewed. From the ground it's often hard to get a clear picture of what has been drawn, but from the air, these massive patterns reveal themselves to be true works of art. Crop circles are very simply constructed in principle, but often fiendishly complex in design.

For nearly twenty years from the late 1970s, there was considerable mystery about what caused these disturbances in farmers' fields. It was speculated that they could be produced by strange weather patterns like freak whirlwinds, or even that they were messages left by alien visitors, either signalling to each other or leaving a sign for human observers to decode. Many agreed that the designs couldn't have been made by human beings without causing more damage to the surrounding crops. But in 1991, a pair of men from the south of England, Doug

8. A cornfield crop circle
(Corcelles-pres-de-Payerne, Switzerland).

Bower and Dave Chorley, admitted that they had been responsible for starting the crop circle craze.

They demonstrated how the circles were constructed, by flattening portions of a crop using one or more planks of wood with ropes attached to the ends. To make the apparently precise alignments, an old hat with a wire loop attached was used to take bearings on landmarks. And because they used no heavy machinery – just a couple of men on foot – there was no disruption to the surrounding crops. With constructions this size, the eye is reasonably forgiving, so the precision is often less than it appears to be.

Many others took up the crop circle gauntlet, producing more and more complex designs. Geometric patterns are still the most common, though they have also been made in the shape of advertising logos and other more worldly images. Despite this, a number of believers cling on to the possibility that some crop circles have extraterrestrial origins. In principle this could be true, just as, in principle, the supermarket at the end of the street could have been built by aliens if you never saw it being built by people – but it's very unlikely. However, the simple origins of crop circles shouldn't take away from their excellence as pieces of short-lived art. Art that uses crops as medium and planks as brushes, but art nonetheless.

On the Nazca plains

There's nothing new in the construction of artworks that are really best seen from the air. Perhaps the most dramatic are the Nazca lines that cross vast swathes of a remote Peruvian desert. These patterns are produced very simply by scraping off the darker layer of small stones on the desert surface to reveal the lighter coloured subsoil. The markings are quite shallow – some as little as 10 centimetres deep, and rarely more than three times that – but the desert has so little rainfall that the lines and the designs they form have remained clearly visible for around 1,500 years.

Many of the Nazca lines are literally just straight lines or simple shapes, but there are also a number of stylized images of animals that have been identified as everything from a monkey to a hummingbird. These are large figures – the biggest between 200 and 250 metres across – while the lines can stretch much further. Like crop circles, some have suggested that the Nazca lines are the work of aliens, or by human beings who have been exposed to aliens and are signalling to them. This is largely because it seems puzzling that such a pattern should be constructed when it can't be seen from the ground. Why make something that can only be seen properly from the air, if you don't have any means of flying? The whole thing seems beyond the capabilities of a relatively undeveloped culture.

This analysis overlooks two factors. One is a natural tendency to use a large scale when undertaking something of a spiritual significance. Just put a medieval cathedral like the remarkable 13th-century Salisbury cathedral, with its 123-metre (404-foot) spire, against the typical domestic architecture of the time. It would be easy to think that the primitive medieval Europeans with their tiny shacks would have been incapable of putting together such a massive structure. But the cathedral was built with the specific intention of constructing something outside of human scale, something its builders thought would be fit for God.

Even without the spiritual imperative, we have a natural tendency to want to go bigger and bigger if the canvas allows it. If you watch people making structures in the sand on the beach, some get surprisingly large. Children will draw lines that extend for many metres, not to signal to aliens, but just for the fun of it. They do it because they can, even though they might not be able to make out their drawings properly without taking to the air. Scale is part of the attraction of having such a large canvas. In a sense, the Nazca lines are quite similar to sand art – it's just that the beach happens to be enormous and will keep the images for hundreds of years, rather than washing them away with every tide.

9. A small section of the Nazca lines, cut through by the Panamerican Highway (Nazca, Peru).

Chalk marks the spot

A third reason for producing large-scale designs can be seen with the smaller but still impressive land-markings that you might see when flying over the UK – structures known as white horses. White horses (although they aren't always horses) are made in areas where the soil is a thin veneer on chalk downs. By removing the turf and soil, a process known as scouring, the artist reveals an area of bright white chalk. Cut the correct shape and you can produce the outline of a creature, whether a horse or a man, or you can cut away a whole area of turf to produce a solid white silhouette image.

Probably the best known of these chalk figures is the white horse at Uffington. In practice, this appears to be a dragon (or, some have suggested, a dog) rather than a horse, but the image, made of a few almost abstract curves, is still very striking. Stunning from the air, a complete view of the carving is difficult to make out when nearby on the ground. But it's still possible to get a feel for what is portrayed, and someone on the ground can walk around it, using the shape as a ceremonial pathway and interacting with it – something that could have happened at Nazca too. These images may well have been constructed with the intention of being walked on.

The Uffington white horse appears to be over 2,000 years old. It might seem impossible to be able to date what is little more than lifted turf, but dating comes from two sources. One is optical stimulated luminescence dating, which relies on the way mineral structures are modified by the impact of natural radiation. These modifications tend to occur in a steady fashion, providing a kind of

10. The Uffington white horse (Oxfordshire, England).

clock when something is buried. But the 'clock' is reset when the minerals are exposed to light. By comparing the chalk on the White Horse and that under adjacent turf, an approximate time for the cutting of the image can be found. At the same time, the distinctive white horse shape appears on a number of Iron Age coins, making it clear that either this carving or an image very like it existed over 2,000 years ago.

In one sense, chalk white horses are like human bodies. The cells in the body are constantly being replaced – even those in bone – so after a number of years nothing remains of the body you had before, even though it's still your body. Similarly, white horses are gradually reclaimed by nature as the turf grows back. They have to be regularly scoured – typically every ten years or so – to maintain the image.

Experiment – How high are you?

This is a technique we will use a number of times to estimate distance in the air. We're going to estimate distances using basic geometry. If two triangles have the same angles, the ratio of the lengths of the equivalent sides is in proportion. If the shortest side of one triangle is twice as big as the shortest side of another, then the same goes for the other pairs of equivalent sides.

Using the same approach, if I know how far away something held at arm's length is, and how big that object is, I can compare it with a distant object. If I know roughly how big that distant object is, I can estimate the distance to it. This is possible because you can consider my arm's length and the height of something I'm holding as two sides of a triangle. The distance to the object outside the aircraft window and that object's height form equivalent sides of a similar but much larger triangle.

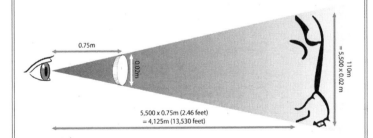

11. Using an object at arm's length to estimate distance.

The starting point of this technique is knowing how far it is from your eye to your fingertips, held at arm's length. Measure it next time you get a chance, but if you don't have a tape measure about your person I can tell you that mine is around 0.75 metres (2.46 feet). You now need a few objects to act as nearby measurements:

- The tip of a ball pen is around 1 millimetre wide (1/1000th of a metre) – 1 metre is 1,000 times bigger than this
- A hole punch hole is around 5 millimetres in diameter – 1 metre is 200 times bigger than this
- A UK penny is around 20 millimetres in diameter – 1 metre is 50 times bigger than this

Measure the diameters of a few of the coins you're most likely to carry to provide further standard measurements.

Now it's simply a matter of seeing how big a known object seems to be. For example, the Uffington white horse is 110 metres long. You can think of cars as being around 4 metres long, a detached house around 10 metres across, and a typical factory/large retail site between 100 and 200 metres across.

Let's imagine you see the Uffington white horse and it looks the same size as a penny held at arm's length. One metre is 50 times bigger than the penny. So at 110 metres across, the horse is 5,500 times bigger than the penny. This means that the distance to the horse is 5,500 times bigger than your eye-to-fingertip distance. So you're around 4,100 metres (13,500 feet) away when you make this comparison.

The traces of the past

It's not exactly difficult to spot something like the Uffington white horse or the Nazca lines if you're on the ground. You might not get the clear image that you see from a passenger's-eye view – but you can't miss these formations if you come across them. The same is true of the floorplan of a large ruin, which might look like random walls from the ground, but can reveal a detailed structure when seen from the air. But from a plane you can also see a range of other human traces on the ground that you wouldn't even be aware of if you were standing next to them.

12. Old Sarum, showing a range of ruins (Wiltshire, England).

Over time, anything from individual buildings to whole towns have been deserted. This can happen when a culture abandons an area, leaving behind, say, iron age settlements or Roman villas. It could be that a source of work dries up when a nearby mine is exhausted, or the land is no longer workable. Initially the abandoned buildings would be obvious ruins like Old Sarum in the picture, but it wasn't uncommon for building materials to be taken and used elsewhere, while relatively insubstantial buildings – often with a lot of wood in their construction – would give way to the ravages of time. Great abbeys and castles may still be very obvious after many hundreds of years of being abandoned, but domestic buildings are likely to end up as foundations that soon disappear beneath the soil.

It's possible to detect such hidden historical and archaeological sites using appropriate instruments. The building materials in the foundations will usually have a different density to the surrounding soil, contain different levels of water, and have other atypical properties that modern detectors can pick up. But there's a much simpler way to spot old sites that you may be able to use while flying relatively low in an aircraft.

The best time to try this out is when you're taking off or landing near to dusk or dawn. At this time, the Sun is low in the sky, so the light hits the ground at a shallow angle. Take a look at open fields, particularly those relatively near towns. Say there was a Roman villa there once, now just a series of wall fragments below the soil, invisible to someone walking across the field. Where there's a solid wall just under the soil, the roots of the plants will be limited and growth will be a little stunted compared with the nearby ground. Where there are ditches, the plants will be a little taller and fuller than usual.

This variation isn't noticeable from the ground. But when the light is coming in at a shallow angle, the taller plants will cast a visible shadow. From the air, this and slight differences in coloration between the stunted plants and the normal ones can make an outline of the building stand out. In an open field, the floorplan of a hidden building can be revealed in shadow form. Whole archaeological sites have been discovered this way. If you're lucky enough to spot a hidden building like this, you'll be seeing a structure that may not have been visible on the ground for a thousand years or more, brought to life by a combination of the plants and the exaggerated shadows of a low Sun. Be careful you aren't confusing the lines where a tractor has driven across a crop with this kind of shadow marking. These will tend to be more open lines rather than closed shapes, and more clearly defined than the subtle differences of light and shadow you're looking for.

Following the water course

As well as the remnants of artificial structures, we can also see the way some natural formations have evolved much more clearly from the air. Rivers and streams are excellent examples. The early stages of a stream's formation are very different from a mature river course. As water drains from slightly higher to slightly lower areas, streams begin to form in a structure that's like the twigs and branches of a tree (the form is known as dendritic – tree-like), though here the shape is working in reverse, as a tree grows out into the branches and twigs, where the water runs in from the twigs into the larger branches, finally feeding the main trunk.

There are two fascinating bits of science visible in the formation of these young streams: self-patterning systems and fractals.

Experiment – Self-patterning systems

You can't do this on the flight – you'll have to wait until you get home. Cover a small tray with wax. This is best achieved by melting the wax and pouring it onto the tray. To get it melted, place the wax in a bowl that's sitting in a pan of boiling water. Try to make an even layer of wax across the tray, then leave the wax to set.

Now take the tray and wedge it at an angle in the sink so water can run down it. Pour a very thin stream of boiling hot water onto the top of the wax in the middle of the tray, so it runs down the slope. (Take care with boiling water!) Initially the water will skitter all over the surface of the wax, but as the wax begins to melt, channels will form in the surface. Once the channels are there, water will tend to run down these pathways. This will melt more of the wax, making the channels deeper and wider. The bigger the channels, the more water can run down them. And so on.

Self-patterning systems, like those formed by hot water on wax, are fascinating because initially there was no particular pattern. The fluid (hot water with our wax, drainage water in the formation of streams) runs chaotically across the surface, influenced by tiny fluctuations in levels. As the fluid flows it starts to eat away at the surface. As soon as a shallow channel has formed, more of the water will tend to flow down that path, reinforcing the initial pattern.

The brain also appears to be a self-patterning system. Its main structure for holding information consists of many millions of special cells called neurons. Each neuron can be linked to hundreds or even thousands of other neurons by tiny filaments called dendrites. It's these connections that seem to represent memories or other stored structures, and initially they are very flimsy. But once they have been made, if they are reused they become thicker. These thicker connections are easier to use – so tend to be used more. Once the initial pattern has emerged, it's self-reinforcing with use.

Fascinating fractals

The second bit of interesting science in stream formation, fractals, is also connected to the chaotic nature of the initial way these structures form. This is not chaos in the sense that newspaper headline writers love, meaning horrendous disorder. 'Chaos' here is used in the mathematical sense. Mathematically chaotic systems are highly dependent on how you start things off. A small initial change can make a big difference in the way the system develops. This is often described as the 'butterfly effect', based on the idea that the flap of a butterfly's wings on one continent could cause a storm on another. This is an over-simplification, but demonstrates the concept well.

Fractals are chaotic geometrical patterns that are 'self-similar'. If you look at the pattern as a whole, then take a fraction of the pattern and blow it up to the same size, that fraction has a very similar structure to the whole. Trees are like this, and so are those tree-like first structures of streams. The fractal form grows because large changes in the direction of the water flows are influenced by tiny variances in the land, the classic recipe for mathematical chaos.

The making of meanders

As the main 'trunk' of the stream or river widens it will, over time, begin to meander. This, again, is a chaotic event where a small change in initial conditions results in a large variation over time. Because of variances in the lie of the land, when a stream forms it's unlikely to be dead straight. If we look at a segment that takes a slight bend to the left as it flows along, the water will have a shorter distance to travel along the left bank of the stream and a slightly longer distance along the right bank.

As the water flows in the stream two things will be happening. If you think of the flow of water around the bend, it will be flowing slightly faster on the left-hand inside of the bend, and slightly slower on the outside. At first sight this might seem the opposite to the way you'd expect. If you imagine a solid object sweeping around a bend, where everything is fixed together, you would expect the inner parts to be slower, because they

13. Well-developed river meanders (Atlana and Koyukuk rivers, Alaska, USA).

have less far to go in the same amount of time. This is why car wheels have to have a differential so the inner wheel on a bend can go slower than the outer wheel. But a stream isn't a solid object, it's a fluid. Different parts of the stream don't have to move together, as they do in a solid object.

The faster movement on the left-hand bank is down to the conservation of angular momentum. Think of ice skaters, spinning around with their arms sticking out. If they bring their arms in, they speed up. Angular momentum depends on the distance a mass is from the centre of spin and the velocity with which it's moving. Such angular momentum is conserved – it stays the same unless you apply a force. When the skater's arms come in, the radius is reduced, so the velocity needs to increase to keep the angular momentum the same.

Similarly, when the water is going around the smaller radius of the inner left bank, it will speed up to preserve angular momentum. One of the results of this is a slightly higher water pressure on the outer bank than on the inner. (You can imagine the water molecules on the inner bank are further apart because they speed up, so the pressure is less.)

This difference in pressure causes a secondary flow to occur, from the outer bank towards the inner, pulling with it material from the outer bank. So soil moves from the outer bank of the curve to the inner bank. This makes the curvature of the bend greater. The curve is bulging out more on the right-hand side, producing a more pronounced bend to the left in the flow of the water.

Where the water returns from the bend to the main stream it goes through a right-hand bend, and a similar effect occurs here, with the right-hand bend also growing over time. The result is a meandering route, twisting first one way and then the other, but not in an even fashion as the flow is influenced by all the tiny starting differences. As the process continues you will often get 'point bars' building up. These are deposits of sand or soil on the inner part of the curve, forming a kind of miniature beach as more and more material is shifted from the outside of the bend to the inside.

Eventually one of these bends can get so sharp that it forms a loop, pinching off a section of the stream that is left isolated as a separate little bend, paralleling the main stream. These are known as oxbow lakes, separate curved stretches of water that run alongside a meander in a stream or river. Often, standing by a river or stream it's difficult to get a full view of the water's progress – but from the air it's much easier to spot the tree-like form of young streams and the meanders of rivers and oxbows that haven't been filled in or modified.

Oxbows are unlikely to be left alone for long in many parts of the world, because surprisingly little landscape is totally natural, without any human influence. There are huge swathes of desert and forest, tundra and wilderness that are mostly unscathed, but anywhere near human habitation – and that means pretty well all of Europe, for example – will have a human hand in its development. Even apparently natural countryside, for instance, is kept the way it is by grazing or other agricultural intervention.

Although there are some regular natural features, it's usually possible to spot from the air the difference between the chaotic formation of the natural countryside and aspects that have been changed under human management. You may be able to see terraces of land, constructed in medieval times or earlier to enable cultivation on slopes, or the two very different styles of field boundaries. The older, small, irregular enclosures tend to be shaped by natural features, while the huge, more linear fields of modern agriculture are designed to make the best use of farming machinery.

How does your town grow?

As you move over habitation, while you're still low enough to see the broad structure of streets, it's interesting to watch out for the differences between evolved and planned towns. Like a branching tree or a dendritic stream, early towns emerged from a collection of lanes that followed natural formations. Often a stream or river, giving communication and water supplies, or the strategic safety of a hill, would be fundamental to the shaping of the settlement. You will find a lot of curves, sudden branches and maze-like structures.

14. An evolved settlement (Wiltshire, England).

Later on, there is a greater tendency to shape the land to fit the town. In a modern town plan you will see rivers and streams funnelled through man-made channels and an increasing appearance of symmetry in the streets. At the extreme there are the regular grid patterns of many American cities, but it can be equally apparent in the symmetrically curved streets of a modern housing estate.

Try to look at a town or a village as a whole object, taking in the shape, treating the roads like veins in a leaf. Does it look like a natural object, or something designed?

15. A planned settlement (Manhattan, USA).

Generally speaking, from the air, the older the settlement, the more organic-looking the visible form it takes. There are some exceptions – some of the better modern suburban developments, for instance, mimic the more natural shapes of an evolved town plan, but even here you can see the hand of the planner in the too-perfect alignments and structures. Architects find it difficult to create fractal chaos, whereas in nature it's the most common form.

The infinite coast

Experiment – Coastal calculations

If you're flying from an island like Britain, you'll soon reach the coast. Take a look at the stretch of coast that's visible. How long would you say it is? Use any identifiable objects – depending on your height, this might be cars, houses, factories – to give scale. As before, as a very rough guide, you can think of cars as around 4 metres long, a detached house around 10 metres across, and a factory or retail building between 100 and 200 metres across.

This is a good chance also to try out your distance estimation technique (see page 59), comparing a known object with something held at arm's length.

16. The complex contours of a rugged coastline
(Provence-Alpes-Côte d'Azur, France).

Your estimate of the length of the stretch of coast you can see from the window is likely to be based on placing as straight a line as you can along its length, but is this a realistic measure? What if you went into every little cove and crinkle in the coastline? Then the distance would be much longer.

This is a small-scale version of the 'How long is the coastline of Britain?' problem. You can define a *minimum* distance for the perimeter of Britain (or any other island) by selecting a ruler size and not measuring any inlets and bumps smaller than this, but as you measure to more and more detail, so the length can be increased indefinitely. In principle, mathematically at least, the coastline could end up being infinitely long.

Experiment – An infinite coastline

For this experiment you need a pen and paper. If you don't have a steady hand, it may be helpful to have something to use as a ruler. Draw a large circle (it doesn't matter if it's not perfect, though you may want to draw around a small plate). Inside the circle with points nearly touching it, draw an equilateral triangle – a triangle with all the sides the same length. Now on the centre of one side of the triangle, pointing outwards, draw another equilateral triangle, one third the size of the first one. (To do this, split the side of your big triangle roughly into thirds, and base the new triangle on the middle third.)

Now do the same thing on one of the outer-facing sides of the second triangle – add a smaller triangle,

---

facing outwards in the middle of the side, one third the size of the new triangle. Carry on this process as long as you like.

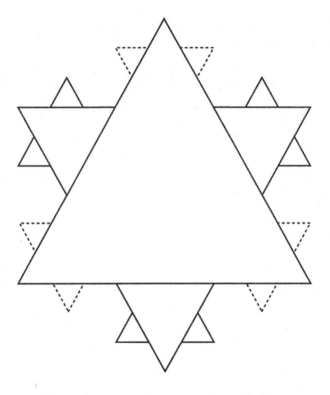

17. The early stages of constructing a Koch curve.

To get the full effect you need to put second-level triangles facing out from each face of the first triangles, then third-level triangles facing out from each of the outer faces of the second-level triangles and so on.

The shape that you are producing is called a Koch curve. It's interesting because it covers a finite area

– it will never come out of the circle – but the distance around the edge of the Koch curve can carry
on growing as you continue to add triangles until it's
infinite in length.

It's a bit like the perimeter of Britain. The distance
around the edge of the original triangle is smaller
than the circumference of the circle, but as we add in
all the crinkly bits it becomes infinitely long.

Like the leaf or tree form of young streams and rivers, the
Koch curve in our experiment is fractal. If you look at a
subset of the curve, it's similar to a larger portion of the
curve. The Koch curve can have an infinitely long perimeter because it's an abstract mathematical shape. It's a
bit different with the coastline of Britain (or your island
of choice). Although in principle you can keep adding
more and more crinkles into your measurement of the
perimeter, this is a physical object made up of atoms.
Eventually, your crinkles would get down to the scale of
atoms and you wouldn't be able to take it any further. So
Britain's coastline is not really infinite – but it can still
be very, very long.

Perhaps what's most interesting here is that the distance is arbitrary. We genuinely can't say how far the
distance is around Britain's coastline. We're used to
science coming up with very specific answers, but this
is a situation where a whole range of answers is equally
correct, depending on what you want to do with them.
There's no single right answer to the distance around the
coast.

Gravity always wins

Those rivers and streams we saw earlier will eventually reach the sea. Children (and some early philosophers) have a tendency to assume that rivers always head for the sea ... because that's what rivers do. This patently isn't true – you can have a river starting fairly near a coast, then heading off inland in the opposite direction to the sea. The reality is that rivers head downhill. They have to – they are powered by gravity. Understanding the force of gravity is key, not only to following the path of rivers, but also in being comfortable with the way that you stay in the air. It might seem straightforward, but we don't always make the right assumptions about forces.

Experiment – Force fields

For this experiment you need to throw a ball into the air. If you're on board a plane, use a scrumpled up piece of paper (but don't throw it too high). Throw it gently into the air so it goes up a few feet, then catch it when it falls back down. Watch the ball as it travels on its flight. Do this a few times. Try to see three separate phases – the ball rising into the air, reaching the top of its motion and falling back towards your hand.

Think about each of those phases of movement. Ignoring air resistance, what direction are the forces acting on the ball just after it has left your hand as it heads upwards, when it reaches the top of its flight, and midway on its journey down?

It might be surprising, but if we ignore air resistance, there's only one identical force acting on the ball in each circumstance. It's the force of gravity, acting downwards. You did provide some upwards force with your hand, but as soon as the ball has left your hand, the only force is gravity downwards. That means the ball is accelerating downwards throughout its journey. As it starts off moving upwards, the result of that acceleration is that it slows down. It's accelerating in the opposite direction to its movement.

At the top of its flight, the ball stops moving. But there's no balance of forces. The force is still accelerating it downwards. Similarly, and perhaps most obviously, on the way down it's still feeling a downward force and accelerating in that direction.

Don't worry if your initial thoughts were different. When this question was put in a survey of science teachers (admittedly, mostly not physicists) the majority got it wrong. It isn't obvious – but when anything is moving, like your plane, understanding the forces involved is essential.

The force of gravity doesn't give us any choices. It always points downwards (or, more accurately, points towards the centre of the Earth). So, driven by gravity, a river heads downwards. Usually the coast is the lowest-lying part of the land locally, which means that eventually the river will find its way to the sea. However, it's entirely possible that there will be a local low point. In that case, the water will accumulate there.

If there's only a small stream flowing into the low point, evaporation and seepage may counteract the rate of flow and so you will end up with a stable lake, fed by the stream. Otherwise, the water level will rise until it tops the lowest of the surrounding areas and a new self-patterning stream will head off towards lower ground.

From river to sea

Many rivers form a delta as they reach the sea. This is a long-term development, where the river has carried small amounts of sediment down to the coast. As the river opens into the sea, the water flows in all directions, slowing down as it moves to the sides. This means that the sediment it has been carrying tends to fall to the bottom, producing a V-shaped protrusion into the sea, carrying the land forward little by little over time to form a delta.

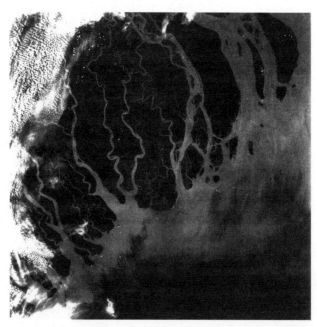

18. The Ganges river delta (Bengal, India and Bangladesh).

Not every river does this. Many have a wider, less well defined mouth, where there's a region of water that's halfway between being sea water and river water. This is an estuary, which is often accompanied by mud flats or wide areas that are covered in water only part of the time.

The distinction between sea water and river water is usually described as being the difference between salt water and fresh water. Certainly if you taste sea water, the general impression is of saltiness. Yet in reality, sea water doesn't contain salt. What it actually contains (among many other things) are sodium ions and chlorine ions. Ions are atoms of an element that have gained or lost electrons. The sodium ion is one electron short, making it positively charged, and the chlorine ion has an extra electron, making it negatively charged.

On its own, sodium is an unstable metal that reacts explosively with water. Pretty well all the sodium on the Earth is locked up in compounds. These are molecules containing different kinds of atoms. (Molecules can also be made up of a single type of atom, such as the hydrogen molecule H_2, which is just two hydrogen atoms joined together.) The sodium ions in the sea have got there as a result of rock material like sodium silicate (a compound of the elements sodium, silicon and oxygen) and sodium carbonate (another compound, made up of sodium, carbon and oxygen) dissolving on a river's journey to the sea, or as sea water passes over and smashes into appropriate rocks.

Chlorine is also a dramatic element on its own. A green-coloured gas, it's poisonous enough to have been used in gas warfare in the First World War (and compounds of chlorine are used to keep swimming pools clean, as it's poisonous to most bugs as well as us). The chlorine ions in the sea have mostly come from underwater volcanoes and vents, which churn out vast quantities of chemicals into the sea. In ordinary sea water, the two sets of ions, sodium and chlorine, are floating around, no more associated with each other than any of the many other ions the sea contains. But if you evaporate sea water, the concentration of the ions increases, the positive sodium is attracted to the negative chlorine and the two join together, forming crystals of sodium chloride or common salt.

Water, water everywhere

If you're flying any great distance, you'll probably spend a good amount of your time over water. A little over 70 per cent of the surface of the Earth – more than two thirds – is covered in water. Looked at from space, the defining feature of the planet is water. Our world is blue with the stuff. In round figures there are 1.4 billion cubic kilometres of water on the Earth. This is such a huge amount, it's difficult to get your head around. A single cubic kilometre (think of it, a cube of water, each side a kilometre long) is 1,000,000,000,000 litres of water.

So why do we get water shortages? Why does agriculture fail across wide areas of Africa because of a lack of water? Divide the amount of water in the world by the number of people and we end up with 0.2 cubic kilometres of water each. More precisely, 212,100,000,000 litres for every person.

If you stacked that up in litre containers, the pile would be around 10 million kilometres high – that's 26 times the distance to the Moon. With a reasonable consumption of 5 litres per person per day, the water in the world would last for 116,219,178 years. And that assumes that we permanently use up water. In practice, much of the water we 'consume' soon becomes available again for future use.

Things are, of course, more complicated than this simple picture suggests. In reality, we don't just get through our five litres a day. The typical Western consumer uses between 5,000 and 10,000 litres. Some is used in taking a bath, watering the lawn or flushing the toilet – but by far the biggest part of our consumption, vastly outweighing personal use, is the water taken up by manufacturing the goods and food that we consume. Just producing the meat for one hamburger can use 3,000 litres, while a 1kg jar of coffee will eat up 20,000 litres in its production.

However, even at 10,000 litres a day, we still should have enough to last us over 57,000 years without adding back any reusable water. So where is the crisis coming from? Although there's plenty of water, most of it is not easy to access. Some is locked up in ice or underground, but by far the greatest majority – around 97 per cent of the water on the planet – is in the oceans, the form you will probably see it in most on your flight.

This isn't particularly difficult to get to, certainly for any country with a coastline, but it is costly to use. The fact that an island nation like Britain is prepared to spend huge amounts of money on reservoirs to collect a relatively tiny amount of fresh water, rather than use the vast quantities of sea that surround it, emphasizes how expensive is the desalination process required to turn sea water into drinkable fresh water. Water shortages are actually power shortages. If we had enough really cheap power we could transport as much water as we liked to the right place, and remove impurities like salt with little effort.

When you're crossing an ocean like the Atlantic or the Pacific it can be very obvious just how much water is out there. For hour after hour, travelling at over 800 kilometres (500 miles) an hour, all you can see is sea. But this isn't a uniform stretch of water. Even from a plane's cruising height you may be able to make out the white peaks of breaking waves and often widely varying colours from the brightest blues through to greens, greys and even yellows. The seas might be immense, but they are not uniform, boring sheets of water.

Time and tide wait for no one

One of the biggest influences on our oceans is the tide. For thousands of years there was confusion over what caused this twice-daily rise and fall of sea level. Galileo, excited by the new-fangled idea that the Earth travelled around the Sun, was convinced that the tides were a side effect of the Earth's motion. He thought that as the Earth hurtled round its orbit and spun around, it pushed the water to one side, rather like a passenger being thrown to one side as a speeding car takes a bend. This was one of Galileo's main arguments for accepting that the Earth moves around the Sun. But there was a slight problem

– his theory predicted only one tide a day, where actually there are two.

Some of Galileo's contemporaries suspected the Moon was responsible for tides, as there was a clear match between the rise and fall of the sea and the position of the Moon in the sky. They suggested that it was moonlight that exerted a strange influence over the water. But this theory had to be discarded when it was pointed out that clouds made no difference to the strength of the tide. Now, we know it's the gravitational pull of the Sun and the Moon that between them is responsible for the tides.

Leaving aside the Sun's effect, which produces the seasonal variations, the Moon does all the hard work. Just imagine the Earth and the Moon, hanging in space. The gravitational attraction of the Earth pulls on the Moon, and the Moon pulls on the Earth. The force of gravity gets weaker the further away you are from the thing doing the attracting, so the side of the Earth that's closer to the Moon feels a stronger pull than the far side of the Earth.

This means that the sea water on the Moon-facing side of the Earth bulges up skyward. As the gravitational pull of the Moon on the far side of the Earth is weaker, the water is less attracted in the Moon's direction, and bulges out away from it. The result is that two high tides are in progress at any one time, one on the side of our planet facing the Moon, the other on the side furthest away from it. These tides sweep around the Earth, following the position of the Moon in the sky.

If the Moon had water on its surface, the equivalent effect would be shockingly powerful. Tides on the Moon, caused by the Earth's gravitational pull, would be all-day

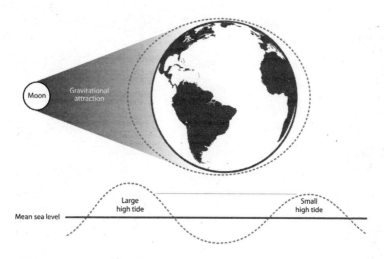

19. The tides produced by the Moon's gravitational attraction.

tsunamis. The Earth is around 80 times more massive than the Moon, and the force of gravity goes up with mass. Double the mass, double the force. So for a similar amount of water, the tidal effects that the Earth would generate on the Moon would be 80 times as large as the tides the Moon generates on Earth.

This may seem too big if you've heard that the gravity on the surface of the Moon is one sixth of that on the surface of the Earth. (Think of those images of astronauts bouncing around on the lunar surface.) How can the Earth's gravitational pull be 80 times larger than the Moon's, but the force of gravity on Earth be only six times bigger?

This is because the gravitational force you feel is altered by both the mass of the body attracting you *and* the square of your distance from its centre. The Moon's mass is 1/80th of the Earth's, but the radius of the Moon is 3.6 times smaller than the Earth's, so on the Moon's surface you're correspondingly closer to the body's centre. This means that if the Moon had the same mass as the Earth, you would feel thirteen (3.6 × 3.6) times as much gravitational force. With 1/80th the mass, you feel 13/80th of the force – around one sixth.

On the crest of a wave

Unlike tides, the waves that we see rippling across the waters are not caused by the Moon, but are solely down to the Sun. It's heat and light from the Sun that powers our weather systems, including the winds, and waves are generated by wind. (The exception here is tsunamis, which are produced by earthquakes and landslips, but the vast majority of waves are produced by wind flowing across the surface of the sea.)

Experiment – Micro waves

Get a glass around three quarters full of liquid. Give the glass a sudden sharp shake (be careful not to splash the person next to you if you're on a flight). The resultant wave is similar to a tsunami. This is a single wave pattern as a result of a shock such as an earthquake, that progresses across the sea, carrying a large volume of water.

Now get your lips close to the edge of the glass and blow gently across the water. You should be able to see small ripples in the water. These are wind waves, like the standard waves on the sea.

If you watch waves, it looks like water is travelling along, but this is misleading. If ordinary waves were truly travelling the way they look to be, they would run much further up the beach than they do, like a tsunami (which does travel). The everyday waves you see at the coast, or in mid-ocean, involve water flowing in a squashed circle, rotating across the top of the wave, down underneath and back up to start again. The *wave* moves forward – the shape moves constantly onward – but the bulk of the water doesn't.

Many of the waves that are so obvious down at sea level can be hard to spot from the air, because they are only a change in shape of a transparent material. Instead, what we can mostly make out is breaking waves, also known as breakers or white horses. These are most common near the shore, as waves are more likely to break when the water is shallow – but you will see them far out in the ocean.

As a wave gets closer to breaking, the height of the wave is getting greater. The higher the wave, the steeper the angle of its front, until the top finally tumbles over, causing the wave to break. This happens in shallow water because, as the wave heads towards the shore, there's less room underneath for the flow of water. As we have seen, the water in the waves is cycling around in a rolling circular motion, and this circle is squashed more and more by the shallowness, forcing the wave top upwards.

The depth of the water also has a strong effect on the direction that waves travel in. Think about waves on a beach. Why do they always head towards the shore, even if the wind is blowing in a different direction? The

20. Waves breaking near the shore (Cape Town, South Africa).

decrease in depth doesn't only change the shape of a wave, it also changes its direction, making it inevitably head inland.

The breaking of waves isn't limited to the shoreline, though. It will happen anywhere the amplitude (the height of the wave) gets big enough. This can happen in mid-ocean if the wind is strong enough, if it's blowing for a long time, and if it's blowing over a wide stretch of water. As the wave breaks, the water goes from flowing smoothly to turbulent flow (we've more to come on turbulence). The collapsing wave peak crashing down on the sea below, and the turbulence, mix quantities of air into the water, generating foam that produces the typical white-crested look we expect of breakers.

What colour is the sea?

The colour of the sea itself varies hugely. This is hardly surprising when you consider the vast variation in what we blithely label as 'sea'. It's as if we called everything from Mount Everest to the Grand Canyon just 'land'. We tend to think of the sea as bland and level – but underneath there's even more variation than we are familiar with on the surface.

When it comes to mountains, for example, the sea has plenty to offer. The longest mountain range in the world is not the Himalayas or the Andes – it's the Mid-ocean Ridge, a continuous chain of mountains under the water running for over 55,000 kilometres through the Atlantic, across the Indian Ocean and up the Pacific off America's west coast. That's more than the distance around the Earth (the chain is anything but a straight line). When it comes to mountains, Mauna Kea on Hawaii has 4,200 metres above the sea, putting it on a par with the higher mountains in the Alps. But if you follow it all the way to its under-sea base it towers up a total of 10,200 metres, dwarfing Everest's 8,800 metres.

However, although Mauna Kea is often labelled the highest mountain on Earth because of this, it really depends on what you consider a mountain to be – because the sea can provide an even greater summit than this. The Mariana trench, the deepest part of the ocean yet discovered, east of the Philippines, is known to plunge at least 11,000 metres – so the edge of this vast underwater chasm rises even higher than Mauna Kea above the trench floor. By comparison, the Grand Canyon, at 1,830 metres in depth, is little more than a crease in the Earth.

Variations in depth are one factor that can influence the coloration of the sea when seen from the air. In shallow water, the colour of the sea bed – which can be anything from brilliant white sand to black volcanic debris – will have a major impact on the brightness of the colour. By the coast we can see brilliant blues, through turquoises and greens to greys. Perhaps the oddest description of the sea's colour to modern eyes is the Ancient Greek writer Homer who called it 'wine dark' – this seems to have been primarily because the Greeks thought of colours in terms of dark and light rather than the specific colours of the spectrum. They had no word for blue, so describing the sea as having a similar shade to red wine did the job.

Another influence over the sea colour is an effect that makes all water slightly blue, so a white sand sea bed takes on a blue tinge. This happens because water molecules are better at absorbing red light energy than the other colours of the spectrum, leaving a blue tint to the remaining light that passes through any water. The sea can also pick up coloration from the sky – a dark grey sky will inevitably result in a dark, threatening-looking sea. And then there's the material floating in the water.

Not every ocean is crystal-clear. Part of the reason why the sea around Britain tends to be a murky grey-green is the suspended mix of biological material – algae, seaweed and more – and fine silt. Not all UK waters are like this, though. There are beaches, for example, on the Outer Hebrides that, as you fly over them, have the same bright coloration as a tropical island.

It's also possible, if you fly over the oceans, that you will come across sections where the colour of the sea owes all too much to human contribution. The biggest rubbish tip in the world is in the Pacific Ocean. Sea currents pull in floating debris to areas either side of Hawaii known as the Western Pacific Garbage Patch and the Eastern Pacific Garbage Patch. Between them, these floating accumulations of debris (they aren't solid enough to call islands) contain a vast amount of material, estimated at well over 10 million tonnes and covering an area greater than Texas. There are similar but smaller patches of rubbish in the North Atlantic and the Indian Ocean.

Above the Clouds

Into the clouds

So far the view has been superb. But at some point on the journey you're likely to pass into cloud. From inside the plane it will look as if it has become foggy outside, but eventually you will break out above the clouds to see an astounding cloudscape around you. In the daytime, up above the clouds, the Sun is always shining and the sky always blue. Below might lie a ruffled eiderdown of clouds, presenting a majestic spectacle as they stretch to the horizon.

There are a number of different cloud types you might notice as you pass through and above them, but before looking at them in detail, we need to find out just what a cloud is. There's always water vapour in the air. We tend to think of the jet of steam that emerges from a kettle as water vapour, but this is misleading. Water vapour is an invisible gas, the gaseous form of water, just as ice is its solid form. What we see as steam is water vapour that has condensed back to a liquid, forming tiny droplets in the air.

We know that water boils at 100°C, so it might seem odd that there's always water vapour in the air, even at room temperature. Just imagine the sea, as the world's main source of water vapour, at normal temperature. What we mean by temperature is a measure of the speed of the molecules in a substance. The faster the molecules move, the higher the temperature. But temperature is a matter of statistics. It doesn't tell us that *every* molecule is moving at the speed you'd expect for this temperature, but rather that *on average* their speed will be the one that corresponds to this temperature.

In practice, some molecules will be moving much slower, some much quicker, than the average. Compared to the water molecules below the surface, the molecules on the surface of the sea that are moving at high speed have a better chance of moving further before they hit another molecule. Some will travel so fast that they can escape from the electromagnetic attraction of the other molecules in the ocean and head off into the atmosphere. If all the molecules in the sea were travelling this fast, it would boil. As it is, it's always losing a few molecules at a time – evaporating – which over the vast surface of all the oceans makes for a whole lot of water vapour.

So water molecules are shooting off as vapour all the time. At the same time, water molecules in the air drop back into the ocean. In any particular set of conditions there will be an equilibrium between water evaporating into vapour and condensing back into liquid. The amount of water vapour in the air is measured by its humidity.

Some of the water molecules will clump together in the air and become tiny droplets of liquid – or higher up, where it's colder, minuscule crystals of ice. These droplets can form due to a temperature change, but are often seeded by floating particles like grains of dust, smoke or pollen. This even happens with bacteria – there are billions of bacteria in the air, which seem to be responsible for producing a lot of this water condensation. The suspended water droplets, like the droplets you see emerging as steam from the kettle, form a cloud.

An obvious question that's rarely answered is, why don't clouds fall out of the sky? After all, water is heavier than air. It doesn't float around if you spill it out of a glass. Of course, water falls from the sky as rain, but why don't the clouds drop out of the sky to land as puddles?

The surprising answer is that clouds do sink. There's nothing magic about them – they have to respond to gravity like everything else. But they sink very, very slowly. It's because the droplets of water are so ridiculously small – as little as 1/100,000,000th of a metre across. At this scale, objects don't behave as we expect them to. Although the drops are subject to exactly the same forces as are visible drops of water, the relative impact of those forces changes.

The force of gravity, dependent on the drop's mass, gets less and less. Meanwhile, the smaller the object, the more influence air resistance will have. Because the droplets are much closer in size to an air molecule than is a raindrop, they are much more battered around by the constant impact of the air. To a droplet of water in a cloud, the air is like very thick treacle is to a small ball bearing. It would take one of our tiny droplets over a year just to drop one metre. In practice, clouds don't stay around long enough to be seen to fall.

Clouds can appear in a range of colours. By default they are white, because they reflect a lot of light – but fuzzily, like ice, rather than directly like a shiny metal. Thinner clouds can pick up colours from the sky, particularly around sunset or sunrise, when red clouds are quite common. Others will be significantly darker, ranging from greys to what appears to be black.

In practice, they will never be darker than grey, but our eye/brain combination can make them look darker than they really are. You can see this effect when you look at a picture of the night sky or space on your TV. They look black – but your screen can't get blacker than it is when the TV is switched off, and that's dark grey. Your brain fools you into thinking the screen is black.

The darkening of clouds happens as the droplets in the cloud join together to form bigger drops. This means that light is less likely to be reflected from the surface, and some colours get absorbed, giving a darker colour. We tend to associate dark clouds with storms because the drops need to accumulate together, getting bigger, before they can fall as rain.

An adventure in cloud-spotting

Clouds are divided by type. These correspond both to the height at which the cloud is located and the shape and density of the cloud. (Technically, the types also include how the cloud moves and changes shape, but we can get away without this for the basics.) These different types are useful for weather forecasting, but also give a lot of pleasure from simply observing them. There are technically a great number of cloud types – around 52 – but for our purposes we can simplify them to ten categories. The original classification identified three families of clouds. These were cirrus (from the Latin for 'hair' – hence wispy, thin clouds), cumulus (meaning a 'heap' or 'pile' for obvious reasons), and stratus (meaning a 'layer' or 'sheet' – again, pretty obvious).

This early structuring was done in 1802 by a pharmacist and amateur meteorologist from London, Luke Howard, and picked up by the likes of landscape painter John Constable, who produced reams of cloud studies. Later, in 1896, the clouds were grouped into nine basic forms, each given a number from 1 to 9. This was later revised to include ten cloud forms – 1 to 10. But the World Meteorological Organization (WMO), the body responsible for the numbering, later changed the range again to be 0 to 9.

This final change of numbering was for a surprisingly romantic reason. The cloud type with the number 9 (which later briefly became 10) was the cumulonimbus. Although this is classified as a low cloud because its base starts well down, the peaks of a giant cumulonimbus climb higher than any other cloud. If you were perched on top of a cumulonimbus you could consider yourself on top of the world – and this is where the expression 'on cloud 9' comes from. The WMO realized they were being spoilsports turning a cloud 9 into a cloud 10, so they reversed their decision.

All the way to cloud 9

Let's take a look at the clouds you're likely to be able to see and recognize from your plane window. The lowest of the low is the stratus. These are the sheets of cloud that can look like a layer of mist starting suddenly above a certain level (mist and fog reach the ground – clouds don't), or a uniform grey blanket of cloud. A variant called stratus fractus is broken into shreds that descend further than the main body of the cloud. Stratus can occur just a few hundred metres over ground level, when you hit cloud almost immediately after take-off.

The second low cloud type – often a favourite when cloud-watching from the ground – is the cumulus. This is the standard cloud portrayed by children making a collage, represented so easily by a clump of cotton wool. With bases around 600 metres and above, cumulus clouds form as a result of thermals, rising columns of warm air that are produced when the Sun heats up the ground. These air columns carry both the water vapour and pollen and bacteria, acting as seeds to trigger cloud formation.

21. A flock of cumulus clouds from above.

Cumulonimbus can start at a similar height, but they go up much further (and have to if they're going to be a cloud 9, as they can also, in a smaller variant, be cloud 3). They can rise as far as 18 kilometres, nearly twice the altitude your plane will cruise at. The top part of a cumulonimbus is more wispy than the bottom, and the whole cloud often forms in an approximate anvil shape, producing the typical 'thunderhead'.

22. A large thunderhead cumulonimbus.

Last of the low clouds is the stratocumulus. One variant of these can form as a cumulus rises and thins out, looking as if that clump of cotton wool has been stretched until it fragmented. But the most common form – in fact the most common type of cloud altogether – makes a broken or striated layer, at its most friendly like a thin sheet of cotton wool that has been stretched until it has lots of gaps, but potentially considerably thicker and more uniform, where it's like a higher stratus with more texture.

23. A stratocumulus layer seen from below.

When we reach the mid-level clouds, it's no surprise that altostratus are like a higher version of stratus, a thin sheet of cloud with few discernible features. They may be thin enough to show the Sun through as a clear shape, or more opaque. The thicker variety is a pretty sure source of rain and has a miserable look to it. It typically lies between 1,000 and 2,000 metres up – still early in the climb of an aircraft through the cloud layers. The thicker variety also comes under a second name of nimbostratus – it switches from altostratus to nimbostratus when it begins to rain. (A nimbus is a rain cloud.)

Altocumulus may be a higher form of cumulus, but it has more visual similarities to stratocumulus, producing shredded sheet effects, sometimes in linear bands.

Finally we reach the high clouds, the ones you're likely to be flying up with (alongside the tops of large cumulonimbus). By the time you reach this height, typically above the 20,000 feet (6,000 metre) mark, clouds are almost entirely made of ice crystals rather than water droplets. You will often meet the cirrus – wispy filaments of cloud that form strands across large areas of sky. Traditionally, cirrus clouds shaped like a comma with a more blobby end have been regarded as a warning of strong winds to come when they stretch out to form long filaments. These 'mares' tails' do often reflect weather that is about to become more unsettled.

Another high cloud, forming over 6 kilometres up like the cirrus, is cirrostratus. This forms more of a sheet than the cirrus, often with clear leading edges. Cirrostratus is often accompanied by cirrus clouds, but covers a significantly greater solid area.

Finally we have the cirrocumulus. Again high in the sky – sometimes as high as 14 kilometres – theses clouds are often striated, with a 'herringbone' pattern. They can be broken up more, looking like very high stratocumulus.

As we've seen, there are plenty of variants of these clouds, plus a range of oddities like mamma – sometimes described as udder-like protuberances that form under stratocumulus and cumulonimbus when there's a strong down-draft. (To see them as mammaries requires a fair amount of imagination – they could equally be described as rounded scales.)

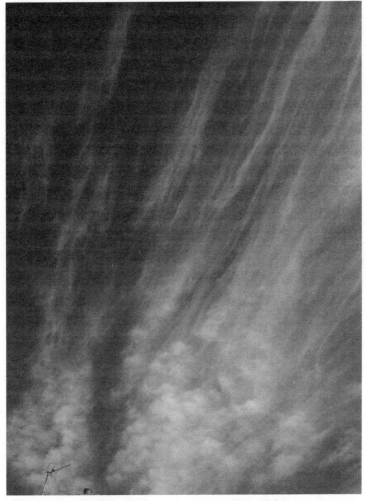

24. Wispy cirrus clouds from below.

No pot of gold for an endless rainbow

Clouds are physical objects, however wispy and insubstantial, but you may also see something from your flight
that doesn't exist at all. Occasionally you may be lucky
enough to see a rainbow from a plane – lucky because it's

a particularly strange phenomenon in the air. Rainbows form when strong enough light from the Sun hits a collection of raindrops – hence the usual requirement for the Sun to be out when it's raining, and for the Sun to be behind you, so it's produced by light that comes over you, hitting a set of raindrops in front of you.

Each raindrop then does a whole collection of the things you might remember from science at school with lenses and prisms. Like a lens, the curved front edge of the raindrop bends the light as it comes in – but it bends different colours by different amounts, so the white(ish) light of the Sun is split into a tiny rainbow. This multicoloured light hits the back of the drop where some of it passes through and some is reflected towards the front. When the light returns through the front of the drop, heading back towards you, it's refracted even more, enough to produce a visible rainbow.

A strong rainbow needs the light to pass through at a fairly precise angle – if the angle between the incoming light from the Sun and the light that heads for your eye is around 42 degrees, you will see a clear spectrum of colours. The raindrops produce a whole ring of rainbow, but this is usually cut off when it hits the ground. In the air, though, you can see the whole thing. The angle is rarely 42 degrees, so the effect tends to be quite faint – but you can see a circular rainbow. If the rainbow is across clouds, which will often be the case, you will usually also see the shadow of the aircraft on the clouds in the middle of the rainbow circle – it's quite a magical effect.

The rainbow doesn't exist – it's a projection of light, not anything tangible, and it joins a range of phenomena that are occasionally seen from planes, and sometimes thought to be unidentified flying objects. There are variants on rainbows called glories, which produce a multi-coloured ring, still with the shadow in the middle, but much smaller than the rainbow, because they are produced by the interaction of light hitting the water droplets from different directions. Often the multi-layered aircraft windows can produce distortions in images, or an apparent floating light that is just the Sun reflected off moving surfaces. There are plenty of optical phenomena that can produce insubstantial but apparently existing objects outside your window.

Over icy seas

We've seen that clouds can be composed of either water droplets or ice crystals. But what about the sea below? Is that always liquid? Everyone will have heard of icebergs, great chunks of freshwater ice that are 'calved' from glaciers, falling into the sea to float there causing hazards to shipping. But sea water itself can freeze over too. It only has to fall below around −1.8°C to freeze, though in practice the movement of the water usually keeps it liquid until it falls considerably below this. However, the ice around the polar regions is mostly frozen sea water.

US science fiction writer Kurt Vonnegut came up with the idea of a very special form of ice, Ice Nine, which appears in his novel *Cat's Cradle*. According to Vonnegut, this Ice Nine was a variant on the ice crystal that was so stable that it melted only at 114° Fahrenheit (45°C). If water ever got into an Ice Nine form, the chances are that under normal weather conditions it would never get out of that form. Should a seed crystal of Ice Nine be dropped into a lake or an ocean it would spread uncontrollably from shore to shore, locking up the water supply and devastating the Earth.

Luckily, Ice Nine doesn't exist (though it was a wonderful concept), although there is a type of ice that forms at very low temperatures with the intentionally similar name of Ice IX. This, however, isn't stable at room temperature, and presents no danger to our water supply.

Up into the sunlight

By now the sea (or land) may have disappeared under a layer of cloud. Unless you're passing through the anvil top of a huge cumulonimbus, the chances are that, if it's daytime, you will see clear blue sky and a brilliant Sun outside.

Experiment NOT to try – Shattering experience

There are a handful of experiments that you definitely should *not* try during your flight. One of these is to discover what would happen if you smashed one of the aircraft windows. If we believe the movies, everyone who wasn't strapped in would be dragged towards the window by an irresistible blast of air and the unfortunate few nearby would shoot out of the plane to a horrible death.

The reality is rather different (though it's still not a good move to try). Firstly, it's very difficult to damage an aircraft window. Protected by a relatively flimsy inner plastic layer, the outer window is very strong.

Should you succeed in making a small hole, the pressure in the cabin could probably be stabilized and certainly wouldn't plunge down explosively. If somehow you managed to take out the whole window, yes, there would be a very rapid decompression. The oxygen masks would deploy and the pilot would drop the plane to a lower altitude which would be survivable but uncomfortable (around 15,000 feet). Smaller items would be sucked out of the window, but it's highly unlikely a human being would.

However, just to emphasize the point – really don't try this. Apart from the discomfort caused in the unlikely event that you succeeded (plus the expense of being charged for the repair bill on a multi-million-dollar airliner), just attempting to break a window is enough to get you arrested, restrained and charged with terrorism offences.

It might seem that there's nothing much to see once you're above the clouds, but there's a surprising amount of science out there.

Voyage to the heart of the Sun

First and foremost, we are in the realm of the Sun. The murky window won't help to give a perfect view, but with less air between you and it, the Sun is the clearest that it will ever be for you. Don't look directly at the Sun – we naturally avoid this for a good reason, as it will cause permanent damage to your eyes. This can happen quickly and even with the Sun partly obscured – every time there's a solar eclipse, hospitals get a stream of patients whose eyes will never totally recover.

The Sun's rays do much more than keep us from eternal darkness, important though that role is. The light from the Sun, which takes around eight minutes to cross the distance between us and our neighbourhood star, brings us the heat that keeps us alive. It powers the weather. And it fuels the plants and algae that underlie our food chain and release the oxygen we breathe. Without the Sun, life would probably not have started on Earth.

It's the Sun, also, that is partly responsible for the blue sky, in combination with air. If you were a flyer in the 20th century, you might have been lucky enough to experience travel on an aircraft where the sky was nearly black in the daytime, demonstrating that it's a combination of the Sun and air that's needed for that over-arching blue. Concorde flew at around 60,000 feet, where the air is significantly thinner, and the sky from the windows was noticeably darker than on an ordinary plane. If you ever have an opportunity to take a Virgin Galactic flight into space, there the sky will be absolutely black, despite the Sun blazing away.

Concorde, incidentally, is a useful lesson for those who say that technology will always advance at an increasing rate, faster and faster. If you look back over the history of travel, initially we were limited to three or four miles per hour on foot, then gained the capabilities of a horse or crude boat. With the 19th century, speeds of at least 50mph (80km/h) became commonplace thanks to the train. Then in the 20th century speeds increased again as air travel eventually took us up to around 500mph (800km/h).

For anyone other than astronauts – which means basically everyone – the maximum possible speed in the 20th century was when travelling on Concorde, reaching a remarkable 1,350mph (2,170km/h), twice the speed of sound. But now we're back to a limit of 500mph. Sometimes technological achievements reach a plateau, at least for a time. There may be supersonic airliners again – the manufacturers are always playing around with ideas – but Concorde's rapid trip from having hundreds of orders to being practically given away shows that the real speed limit is not so much technology as political will.

Why is the sky blue?

In the daytime, assuming you aren't on Virgin Galactic, you're likely to see a blue sky outside your aircraft window. 'Why is the sky blue?' is one of those questions that most children ask at some point – and often they get answers that are a little way from the truth. It's not, as some have suggested, a reflection of the blue sea. In Victorian times it was thought that dust and other particles in the sky gave it the blue tinge – but in fact the source is more subtle, a direct interaction between the air molecules and incoming sunlight.

Visible light comes in a range of colours, which we see in the spectrum of the rainbow from red through to violet. The traditional seven colour distinctions (red, orange, yellow, green, blue, indigo, violet) in the rainbow are arbitrary, an arrangement devised by Isaac Newton.

If you look at a full spectrum of light, you might either think it has millions of subtly different colours, or five or six broad bands of colour – hardly anyone can see seven colours in a rainbow.

We aren't entirely sure why Newton came up with the number seven, including those obscure shades indigo and violet, but there's a strong feeling that he was drawing a parallel with music. In the musical 'spectrum' there are seven notes, A to G, before completing the octave and returning to the next A up. Newton, it's thought, felt that there also ought to be seven colours in the visible spectrum and forced that strange set on us. Surprisingly, the spectrum couldn't have been named this way just 100 years earlier. 'Orange' became the name of a colour only a few years before Newton was born. Until then it was just the name of a fruit.

Newton showed (as does the rainbow) that the light from the Sun contains the whole spectrum of colours (and a good few we can't see as well). It was thought in Newton's day that the rainbow colours were produced by imperfections in the glass of a prism, tinting the white light. But Newton split off a single colour and sent it through a second prism, which didn't change the coloration – the prism wasn't tinting the light. He also found he could recombine the colours to make white light. The colours are all present in sunlight.

When this light containing a mix of photons with different energies (which we see as colours) passes through the air, some of the photons that make up the light get absorbed by the electrons in the gas molecules of the air. Soon after they are re-emitted and shoot off, but in all directions. This 'scattering' means that photons are flung across the sky, no longer all on the same path.

The scattering effect is different depending on the energy of the photons. Higher-energy photons (towards the blue end of the colour spectrum) are more likely to be scattered. This means that away from the Sun itself, the light being scattered from the air molecules has a blue tinge, making the sky blue. Because the red end of the spectrum is scattered less, when sunlight is passing through a lot of air, which happens when the Sun is low in the sky, the light that isn't scattered, coming direct from the Sun, takes on a red tinge – this is why the setting and rising Sun is so red.

Why does the Sun keep shining?

When a child draws the Sun it will be yellow, but in reality, seen without the interference of the atmosphere, the Sun's light is white. Confusingly, it *is* classed as a yellow star. This classification is because the yellow component of the light the Sun gives out is the strongest, but it does provide the whole spectrum of light, as can be seen when the light is shone through a prism, just as Newton did, breaking it down to its component colours.

The natural colour of the Sun reflects its surface temperature of around 5,500°C. If this seems rather cool, the temperature rises to around 15 million°C at the Sun's core. We're dealing with a massive object here – the Sun is around 1.4 million kilometres across, over 100 times the diameter of the Earth, and weighs around 2,000,000,000,000,000,000,000,000,000 tonnes, a third of a million times the weight of the Earth.

For a long time it was a puzzle how the Sun managed to keep burning. Victorian scientists did calculations on how long a ball of coal that size could remain alight and reckoned that the Sun could stay active for only a few million years, nowhere near long enough for the kind of age that geological discoveries were suggesting for the Earth. It was only with an understanding of quantum theory, the science of very small things like atoms, that a practical mechanism for the Sun emerged, one that has enabled it to stay in action for the last 4.5 billion years – around half its predicted active life.

The Sun works by a process called nuclear fusion. It combines the nuclei (the central core of atoms) of the smallest element, hydrogen, to build the next biggest element, helium. (Helium has that name because it was first discovered in the Sun – *helios* being the Greek word for Sun – before we found enough on Earth to fill party balloons and make silly squeaky voices by breathing it.) In the process of fusion, a small amount of mass is converted to energy. Thanks to Einstein we know that the conversion of mass to energy is described by the simple formula $E=mc^2$. The 'c' here is the speed of light, which is huge, so squaring it means you get an awful lot of energy from a tiny amount of mass. Just one kilogram of mass converting to energy produces the equivalent of a large power station running for around six years.

The Sun dwarfs anything we can ever contemplate doing in energy production. Each second, around 4 million tonnes of matter is converted to energy – that's 24 billion years of running a power station. Or to put it another way, the equivalent of 756,000 trillion power stations operating simultaneously. If we covered the entire surface area of the Earth (land and sea) with this number of power stations, each power station would be less than a centimetre across.

Most of that energy that the Sun constantly pumps out heads off into space, but a small fraction hits the Earth and keeps us alive. Just 89 billion megawatts reaches the Earth, less than a billionth of the Sun's output – yet that's still 5,000 times the current total energy consumption for our whole planet. The strange thing about the Sun is that when scientists first realized how it worked and looked in detail at the processes that keep it in action, they discovered something bizarre. Even with nuclear fusion, it seems that the Sun shouldn't work at all.

The reason for this is that those nuclei of hydrogen have to be squeezed very close together in order to fuse. Very, very close. But the nuclei have a positive electrical charge. They repel each other like two magnets do if you bring the same poles together. And this repulsive force gets bigger and bigger as the nuclei close in on each other. Even the immense temperature and pressure at the heart of the Sun isn't enough to get the hydrogen nuclei close enough to fuse. The Sun should be a damp squib.

Taking a trip through a quantum tunnel

The reason the Sun does work is down to the strangeness of quantum physics. When we look at very small things like atoms, they don't behave like objects on the scale we're familiar with. For example, until you make a measurement and locate it exactly, an atom doesn't exist in just one place, like a ball or a table. Instead it has a range of probabilities of being anywhere in the universe. The further away from the centre of this range, the less likely we are to find it – but it really could be anywhere.

This means that if you put an atom in a box, there's a small but real probability that it will jump right through the box and appear on the other side, when you take a look and find out exactly where it is. It's as if you parked your car in a garage and came along next morning to find it had jumped through the garage wall and appeared in the drive, all on its own. (Given that your car is made up of quantum particles, it could in principle do this, but the chances of all the atoms making the same jump at the same time are so low that it isn't going to happen in the lifetime of the universe.)

This process of getting through a barrier like a wall and appearing on the other side is called quantum mechanical tunnelling. Mostly the hydrogen nuclei in the Sun stay roughly where you would expect them to be, but occasionally one will tunnel through the barrier caused by the repulsion that keeps them away from other nuclei, and will appear so near a neighbour that fusion takes place. Although this is very unlikely for any particular hydrogen nucleus, there are so many of them in the Sun that statistically a vast number will succeed, resulting in a steady transformation of hydrogen into helium.

The Sun is more obvious above the clouds than it is from down on the ground – but we still tend to underestimate just how amazing it is. As we've seen, this vast nuclear furnace in the sky is responsible for life. (In fact the Earth itself wouldn't exist at all without the Sun, as the star's gravitational pull was responsible for the Earth forming in the first place.)

The Sun is around 150 million kilometres away. This means that light from it takes around eight minutes to reach us, as does gravity, which we now believe to be carried by a stream of particles called gravitons, just as light travels in the form of photons. If, somehow, the Sun disappeared, for eight minutes we wouldn't realize anything had happened. It would still appear to be there as the remainder of the photons and gravitons crossed space to us. Only after the eight minutes had elapsed would we lose its light and gravitational pull.

Crossing flight paths

You may well see another plane at some point on your flight. There are strict rules on the separation of planes in flight, but they often travel in air lanes (also known as airways) to make air traffic control simpler, which could well result in spotting another aircraft. Although the rules for separation vary depending on conditions and the height you're flying at, you would usually expect planes that are relatively close to each other sideways to be separated by at least 300 metres (1,000 feet) in height on the ascent, and at least 600 metres (2,000 feet) in height at cruising altitudes.

This doesn't apply if the planes are at least 3 to 5 miles (5 to 8 kilometres) apart. With that kind of vertical separation they can be on the same flight level.

Experiment – How far away is that plane?

Next time you see another plane, try out the distance estimation technique (see page 59). An airliner is typically between 30 and 70 metres long. Let's assume it's 50 metres. Say your arm's length measurement of the plane you see is four ball pen points – around 4 millimetres. There are 250 of these in a metre, and 12,500 in 50 metres. So the distance is around 12,500 times your measured distance to arm's length. That's 12,500 times 0.75 metres, or 9,375m – around 9 kilometres. Of course that length was an estimate – the limits I gave of 30 and 70 metres would make the plane between 5.6 kilometres and 13 kilometres away.

Although separation rules should generally keep aircraft away from each other, modern aircraft have collision avoidance systems on board to doubly reinforce the rules of the air lanes. Airliners are usually equipped with a traffic alert and collision avoidance system (TCAS). The TCAS equipment sends out requests for information to other aircraft fitted with the same technology. Each aircraft has built into it a device called a transponder. This is a radio-activated automatic radio transmitter. So when it gets a request from another aircraft, the transponder springs into life, broadcasting its location. That way the other aircraft can build up a picture of the positions of any other aircraft around and warn the pilot of any potential threat long before it becomes real.

Leaving a trail in the sky

Whether or not you see another aircraft – and it's entirely possible to have a whole flight without seeing another plane away from the airports – you may well be able to see where another aircraft has been. If you look up at the sky from the ground you will often see what look like very straight, very thin clouds, crossing the sky like an aerial roadway.

25. Contrails: artificial clouds produced by condensation from jet engines.

These are contrails, trails of water droplets left behind by a plane. When you yourself are up in a plane you might cut across a contrail or come very close to one, able to see the route the plane took without knowing for certain (unless the plane is in sight) which way it went. The best visual clue to direction is that a contrail tends to fan out over time, so the end nearest the aircraft will be tighter together, still potentially having more than one trail from the different engines, which will merge further down the trail.

'Contrail' is just a collapsed version of the term 'condensation trail', the US equivalent to what used to be known as vapour trails in the UK. Of the two terms, 'condensation trail' is the more accurate description. You can't see water vapour, the gaseous form of water – it's entirely transparent. It's only when droplets of water or tiny ice crystals form that the trail becomes visible. A contrail is, simply put, an artificial, strangely shaped cloud.

To understand why contrails form, we have to go back to the way an aircraft engine works. Inside the combustion chamber, aviation fuel is burned. Leaving aside any impurities, this fuel is a mix of hydrocarbons. Although hydrocarbons can be big molecules, they are made up of just two basic atomic building blocks – hydrogen and carbon (hence the name hydrocarbon). When a hydrocarbon burns, the atoms are combined with oxygen atoms from the air. What we think of as 'burning' is really just a chemical reaction, involving combination of a substance with oxygen, that throws out heat. The carbon atoms get together with oxygen to make the greenhouse gas carbon dioxide, and the hydrogen atoms join up in pairs with an oxygen atom to form H_2O – water.

Because of the high temperature of combustion, that water emerges in the form of a gas, but as it's expelled from the back of the jet and meets the cold air, the vapour will condense to form tiny droplets or (if it's cold enough) ice crystals. There are some limits on contrails. They don't stream from the jet engines directly, as it takes time for the gas to cool enough to form droplets, so they form some distance behind the plane. And they rarely appear below 10,000 feet, as the air gets colder the higher you are, and the temperature needs to be low enough to rapidly cool the vapour, before it's dispersed.

In principle there are contrails emerging from each engine, and you can see this if you're close enough to the aircraft producing one. But before long, they will merge, first into one per wing (if the plane has four engines) and then one for the whole aircraft, the form in which they are most likely to be seen from the ground. When the aircraft is at a low altitude – far lower than the height where contrails form – you may see what looks like a very thin contrail emerging from the wingtip of your plane. Sometimes these have caused panics on board when passengers thought the aircraft was on fire. But what you're seeing is neither smoke nor a conventional contrail.

Instead, this is a visual effect of the wingtip vortices, the spiralling trail of turbulence often produced by aircraft wings that is the main reason for the length of separation between planes on take-off and landing. Although by no means as common as contrails, these vortex trails are formed when the drop in pressure and temperature within the vortex causes water vapour to condense. This isn't water vapour produced by the plane's engines, as is the case with a contrail, but the natural humidity in the air, which is usually higher at low altitudes.

Is there life out there?

You might think that another aircraft is the only way you'll see life outside your window at altitude, but there's a range of possibilities for life to exist on the other side of the glass. The most prevalent we have already met – bacteria. As well as acting as seeds for the formation of water droplets in clouds, bacteria are light enough to be carried many miles on the air currents. As many as 1,800 different types of bacteria have been detected in the air above cities, while bacteria have been detected above 20,000 metres up, twice the typical cruising height of a plane.

A few insects too can scale the heights. It's generally thought that a bumblebee holds the altitude record for insects. These creatures live at least 18,000 feet up on Everest, and have been proved capable of flying around in lab conditions up to around 30,000 feet. They may not stray far from the mountains, though, so you're unlikely to see one buzz past your window, even if it could cope with the turbulence caused by the passage of the plane.

You're more likely to see birds at some points on your flight. Most songbirds stay below 2,000 feet and waterbirds around 4,000 feet, but some can go much higher. The bar-headed goose, generally considered the highest-flying of the migrant birds, makes use of the jet stream, some travelling as much as 1,000 miles in a single day. To do this they have to get up to as high as 30,000 feet. To keep going with such limited air supplies, the geese combine the standard bird breathing system, which circulates air through the lungs twice to get extra oxygen, with a special form of haemoglobin that's particularly efficient at latching on to oxygen.

The geese aren't the only high-flyers. Whooper swans have been sighted from aircraft at 27,000 feet, and even mallards have been sighted over the 20,000-foot mark. Remarkably, it's a vulture that holds the avian world altitude record. A Ruppell's griffon, a bird with a sizeable 3-metre wingspan, was detected the hard way at nearly 38,000 feet over the Ivory Coast. The bird, unfortunately, perished when it was sucked into a jet engine.

Going walkabout

By the time you have reached your cruising height, the captain will probably have switched off the 'fasten seat belts' signs so you can take a walk around the cabin. On a long flight this is a good thing to do on a regular basis to reduce the chances of deep vein thrombosis (see below). Your chances of taking a wander around may be limited by the cabin crew's progress down the aisles with food or sales trolleys, but do take the opportunity, partly for your health and partly to take in some of the details of the plane.

Experiment NOT to try – The doorway to disaster

If you're wandering around the plane, you may well find yourself by one of the doors. There's a horrible fascination when faced with the handle that opens the door. That is, after all, the only thing between you and the outside. It's hard not to think, 'What would happen if I opened the door?'

Some people think that, as on trains, the doors are locked on departure. You will often hear a command to the cabin crew along the lines of, 'Doors to automatic. Cross check', and it may sound as if the doors are being locked. In fact this instruction is to put the inflatable evacuation slides onto automatic, so they shoot out if the door is opened. (The 'cross check' part just means to check that the door across the way is also set correctly.) No one seems quite sure why this can't be centrally controlled.

In practice the doors on a plane don't need to be locked. If you watch an aircraft door being opened it swings in an unusual way. This is because it first has to be opened inwards before manoeuvring it out of the way. Once the plane has taken off, a significant pressure difference soon builds up between the inside of the plane and the outside, as the cabin pressure becomes higher than the rapidly falling outside pressure. This air pressure difference forces the door into place. To open it you would have to pull against the air pressure, well beyond the capabilities of human muscles.

However, it's very important *not* to test this out. Leaving aside how embarrassing it would be if I got it wrong and you did succeed in opening the door, like trying to break a window, any attempt to tamper with the door handle in flight will be regarded as a dangerous act. The crew will restrain you, probably using plastic loop handcuffs, and will radio ahead to have police standing by so that you're immediately arrested on arrival. Don't try it.

Even if the trolleys aren't blocking the aisles, all too often those 'fasten seat belts' lights come back on before you have a chance to get moving, and you're encouraged to return to your seats. You're about to encounter turbulence, the air industry's answer to a theme park roller coaster.

Travelling through bumpy air

Turbulence sounds more complicated than it is. It's just a sudden change in the movement of the air around the plane, causing a jerky motion that can be anything from the feeling of travelling along a bumpy road to a sudden plunge that leaves your stomach floating in mid-air. There are a number of possible causes for this interference with a smooth flight. It can be precipitated by a change in temperature in the air, or wind shear where two sections of the air are moving in different directions. It can be invisible in clear air, or all too obvious around a storm. The essential outcome is that as the aircraft moves forward, it suddenly encounters dramatic changes of movement in the air around it, resulting in that bumping or a sudden and unnerving drop.

Because turbulence can be a disconcerting experience, it's important to realize that no modern airliners have *ever* been brought down by turbulence. It isn't going to cause a crash. People have been injured and even killed as a result of turbulence, but this is because they haven't fastened their seatbelts and have hit their heads on the ceiling – or because poorly secured luggage in overhead lockers has fallen out. Take turbulence seriously and it

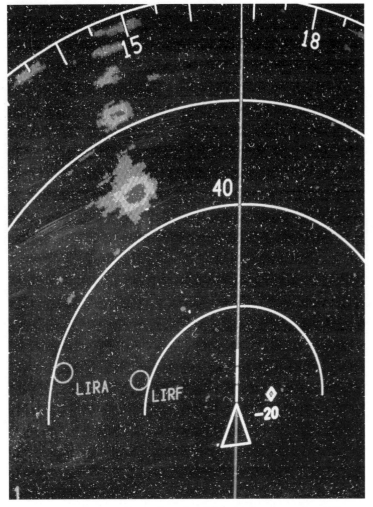

26. A storm on an aircraft radar (this is the cumulonimbus shown on page 97).

won't be a problem, so you really can sit back and enjoy the ride.

If the turbulence is caused by a storm, it's likely to occur while the aircraft is climbing, or heading down towards an airport. The cruising height of airliners is above most storm clouds (this is why airsickness used to be so much more common in early unpressurized planes, because they couldn't fly so high). As we've seen, the thunderheads of some cumulonimbus clouds can extend far above 40,000 feet (12 kilometres) so can tower above even a cruising aircraft, but unlike clear-air turbulence, such storms are both visible and show up on the plane's radar, making it easy for the flight crew to swing around and avoid the worst of it.

Even so, if unavoidable, you may have to pass through storm clouds particularly in climb and descent, which can be worrying if there's thunder and lightning outside your window. Lightning is an incredibly powerful natural phenomenon, and it's only reasonable to be a little concerned about having it frying the air just a few feet away, or even striking the plane. The good news is that it's rarely a problem. But first we need to establish just what thunder and lightning are.

The flash of lightning

Thunder and lightning are not separate things – thunder is just the sound that lightning makes. We tend to think of them as being separate because the thunder usually lags seconds behind the lightning flash. This reflects just how much faster light is than sound. Say a thunderstorm is 10 kilometres away. The flash of light, travelling at 300,000 kilometres per second, will take just 1/30,000th of a second to reach you – effectively you will see it as soon as it happens. The sound, lumbering along at a relatively slow 340 metres per second, will take over 29 seconds to arrive.

Many of us were taught to tell how far away a storm was by counting the seconds between the flash and hearing the rumble of thunder. I was told three seconds per mile, but in fact sound travels only 0.21 miles in a second, so it takes nearly five seconds for it to travel a mile (a count of three is more like a kilometre).

Lightning is nothing more or less than a discharge of electricity. Air is quite a good insulator (meaning that it resists the flow of an electric charge), but if you crank up the voltage of the electricity (see page 126), you will eventually get a spark to break down the resistance and jump through the air. A rough guide is that at normal humidity (wet air conducts better than dry) it takes around 30,000 volts to get a spark to jump across 1 centimetre. To make a metre-long electrical discharge under normal circumstances would take around 3 million volts.

A static charge

Exactly how lightning forms is still a little uncertain, but what is clear is that it starts with static electricity.

Experiment – Static pick up

Static electricity is the build-up of electrical charge on an object. It typically results from an object either accumulating negatively-charged electrons or losing them to become positively-charged.

If you have a plastic item like a comb or a pen, rubbing it on your hair will give it a static charge. Tear off a very small piece of paper (say half your little fingernail in size) and place it on your table or your lap.

Rub the plastic object vigorously for ten to twenty seconds in your hair, then hold it over the paper and bring it slowly towards it. The paper should jump up and stick to the plastic object well before they come into contact with each other. This is static electricity showing its hidden power.

When you rub a plastic object on your hair it gains electrons, the tiny negatively-charged particles that hover around the outside of atoms. The plastic becomes negatively-charged, while your hair gets a positive charge. This is a direct physical effect. The electrons are, effectively, rubbed off your hair, attracted to the surface of the plastic. When you then bring the plastic object near a piece of paper, the negative charge in the object pushes electrons away from the surface of the paper. This leaves the top of the paper positively-charged. So you've got a negatively-charged piece of plastic and a positively-charged paper. The positive and negative attract each other and the paper jumps up. This works with any size paper, but the force isn't particularly strong, so you need a small piece to get it to work well.

Static electricity is often made by rubbing one thing against another. The production of electricity this way is called triboelectric, combining the Greek for rubbing (*tribos*) with electricity. The best known static electricity generators, called Van der Graaf generators, work this way. A belt rubs against rollers, producing a static charge that is then built up on a metal dome. These devices can produce millions of volts, though high-voltage 'artificial

27. Forked night-time cloud-to-ground lightning (USA).

lightning' displays are more often produced using a Tesla coil, which is a form of transformer that can push voltages up to very high levels.

Making lightning

Although we aren't 100 per cent sure why lightning forms, it appears to be due to the movement of water drops and ice crystals in the air causing an electrical charge, possibly due to friction – tiny triboelectric charges, but spread across so many drops that it adds up to a massive amount – or it could be the interaction between the natural electric field of the Earth and moving droplets. The sheer size of a thundercloud means that accumulation of the tiny charges on all the drops can build to extraordinary levels.

This very powerful negative charge in the cloud has the same effect as the plastic has on the paper when you pick it up with static electricity. The highly negatively-charged cloud induces a positive charge in either a nearby cloud or the Earth. (Lightning can go from cloud to cloud or between the cloud and the Earth.) The repulsion of the negative charge in the cloud pushes away electrons in the other cloud or the ground, leaving the nearest surface positively charged. Then comes the strangest thing about lightning.

A relatively weak electrical discharge runs between the negatively-charged storm cloud and the positive charge it has induced – say from the cloud down to Earth. This discharge ionizes the air. Ionization means it strips electrons off atoms, leaving the very atoms of the air charged and making it more conductive. This first stroke, called a leader, has set up a pathway. Now the main discharge, the one we see, called the return stroke, takes place. This goes in the opposite direction – if the lightning is striking the Earth, the return stroke runs from the ground up to the cloud, the opposite direction to the one we would expect.

Electricity on the move

Once the electricity is flowing, we've moved from static electricity to current electricity – it's electricity that moves. This is where those basic electrical terms you probably learned (and forgot) at school come in. Voltage is the electrical equivalent of force. It's the amount of 'push' on the electrons, the tiny particles that surround atoms and that carry electricity. Once those electrons start to flow, the amount of current flowing is measured in amps. The terms used in electricity are so familiar that we don't always notice that words like 'current' and 'flow' were taken straight from water systems. But electricity is a bit different from water. (Luckily, or we would have to block up electrical sockets to stop it leaking out.)

When the terminology of electricity was put together it wasn't clear what was happening. Michael Faraday and the other scientists of the time didn't even know that atoms existed, let alone electrons. They knew that something – 'electric current' – was flowing, and arbitrarily decided which way it went, sending it from what they called the positive terminal to the negative. In fact, when electrons were discovered it was found that they – the real thing that flows – moved in the opposite direction to the way electric current is labelled – but it was too late to do anything about it.

Once you have a flow of electricity you produce power. Power is just the speed at which work can be done. It's the rate at which energy is transferred from place to place. Just as was the case with engines, power is measured in watts, and for electricity it's just the voltage times the current – the amount of push times the amount of electricity flowing. And the final part of our quick trip into electrical terms is joules. As we've already seen, these are the units of energy, whether it's electrical energy, the energy that makes a car go, or the energy content in food. We still measure food energy in calories, which is an older unit of energy (and confusingly a food Calorie is actually 1,000 calories, or a kilocalorie) – but joules are the standard unit. A joule is just a watt for a second. So each second, a 100-watt light bulb uses up 100 joules of energy.

The power involved in lightning can be immense. Where something like a Van der Graaf generator makes millions of volts, it operates only at a tiny fraction of an amp, so the power – volts times amps – comes out as a small value. In a lightning strike, there's a high voltage *and* the flow can be more like 30,000 amps, producing as much as half a billion joules of energy – the total output of a good-sized power station for a second.

When this amount of power rips through the air, the result is that the air molecules start to move extremely quickly – the temperature shoots up to as much as 20,000°C, significantly hotter than the surface of the Sun. This sudden change in temperature produces a pressure wave, a shock wave that is, in effect, an explosion; and it's that pressure wave ripping through the air that we hear as thunder.

There's safety in metal boxes

It isn't uncommon for lightning to strike an aircraft (although pilots will avoid storms if at all possible). This doesn't mean that the passengers are going to be fried, though. If you strip away all the fancy bits, an aircraft is a metal box. Early on in the exploration of electricity, Michael Faraday discovered that you can't get an electric charge through to the inside of a metal box (or a metal mesh – it can have holes in it, so windows aren't a problem). As the electric charge builds up outside such a box, called a Faraday cage, the electrons in the metal that makes up the cage move around to cancel out any charge on the inside. The result is nothing gets through the cage. This is why you'll often see it said (correctly) that one of the safest places to be on the ground during a thunderstorm is inside a car. And the same goes for a plane.

The plane itself, however, does need some protection. The biggest concern is that the surges in voltage in the skin of the plane will disrupt the electrical systems essential to the control of the aircraft. Although the electric discharge itself can't get into the plane's shell, it can produce electromagnetic effects that can induce currents, like a transformer in action. Because of this, all airliners have built-in lightning protection systems that discharge any electrical build-up – a modern version of a lightning rod on a church tower.

Grounded by the ash

As we've been looking at some hazards of flying, even though you're very unlikely to encounter them directly, it's worth also taking a look at the stuff that stopped planes flying over much of Europe in 2010. It was all the fault of a volcano that few people had heard of before. In fact, hardly anyone could even pronounce it – the Icelandic volcano Eyjafjallajökull (it's *very* roughly ay-va-there-vtl).

Aviation was pretty well shut down over the whole of northern Europe from 15 to 23 April 2010, with further intermittent closures through to mid-May. At the time there were dire warnings that this could go on for months. The effect was to produce huge travel disruption. Up to then, the assumption was that you can always hop onto a plane and fly anywhere, but many millions of people saw their airspace closed and had to resort to road, trains and boats – or not travel at all. Not everyone was upset, though. Skies that were usually constantly filled with contrails were suddenly pristine. Anyone who lived near an airport could sit out in their garden and enjoy the spring weather.

Volcanic eruption

You're unlikely to be flying anywhere near an active volcano, so you probably won't see one from the air, but a volcano could be responsible for one of your future flights being cancelled or diverted, and volcanoes are without doubt a force of nature to be reckoned with. They are always with us – at any one time there are around 1,500 volcanoes that present some sort of threat, though many are restricted in their impact to the familiar dangers of lava flows and covering a region in ash, as happened at Pompeii and Herculaneum in Italy when Mount Vesuvius erupted in AD 79.

28. The active Pico do Fogo volcano (Ilha do Fogo, Cape Verde).

Some volcanoes, though, pour clouds of very fine ash into the sky. Apart from producing irritating dust layers on your car, if there's enough of this ash, thrown out with enough force, it can spread around the world. When Krakatoa erupted in 1883 around 20 cubic kilometres of ash and rock was spewed out by the explosion – the equivalent of a cube of material around 20 kilometres on each side.

This ash was thrown up 80 kilometres into the atmosphere and travelled around the Earth, following the shock wave from the eruption which was measured passing around the globe a total of seven times. When the dark ash suspended in the atmosphere got in the way of sunlight, it reduced global temperatures by over a degree Celsius and disrupted weather patterns for several years. The ash was detected everywhere around the world.

In 1883 there was no air travel, but today such an eruption would shut down aviation worldwide, potentially for at least a year. The problem is with the very fine ash. In the case of Eyjafjallajökull, this was particularly bad because of Iceland's starkly contrasting combination of cold temperatures and very hot volcanoes. Cold water, particularly from the glacial ice above the volcano, rapidly cooled the blazing hot lava, producing masses of tiny glass-like splinters, which were then thrown up into the air by the explosive force of the eruption.

Such particles of ash are too small to do much damage directly, but if a lot of this kind of ash gets into a jet engine it can melt again, stick to parts of the engine and solidify, causing problems and, at the extreme, engine shutdown. In the most famous case, British Airways

Flight 9 in 1982, all four engines on a 747 stopped after it passed through a volcanic ash cloud. Luckily, once the engines cool a little, the ash becomes easily fractured, so the engines can be restarted, but in the case of Flight 9 it was only after twelve minutes without engines, unpleasant to say the least for all on board, with the pilot regaining control after losing 25,000 feet in altitude.

Airlines and regulators have learned a lot since this incident. Now, not only as we saw in 2010 is every precaution taken to ensure that aircraft don't fly into volcanic ash clouds, but also volcanoes are monitored to ensure there is plenty of warning of hazards. There's no risk of a sudden ash cloud taking an aircraft by surprise any more. It might be irritating not being able to fly in such circumstances – and it's almost certainly true that the authorities were over-cautious in 2010 – but we do still have to make allowances for nature. Volcanoes remain beyond our control.

In the radiation zone

With no volcanoes to watch from your window, and with any storms on your route bypassed, the last remaining concern you might have from the outside of the plane is radiation. This doesn't put the aircraft in any danger, but it's something you need to be aware of, if only to put the risks from body scanners into perspective. It's a hazard of flying that's rarely mentioned, but it does exist. It's important to know up-front, though, that you're in no danger right now. Taking a single flight has very little effect. This is a cumulative problem if you do a huge amount of flying.

Radiation is a word that's often misused. You may have seen scare stories about radiation from phone masts or wireless networks. The use of the word here is little more than propaganda. Technically what phone masts and wireless networks produce *is* radiation, but this has nothing to do with nuclear power and atomic bombs. What they broadcast is electromagnetic radiation. That's just another term for light. As we've already seen looking at radar and X-rays, light has a huge spectrum, with the familiar visible light colours from red to violet sitting roughly in the middle, and other invisible types of light stretching far off in both directions.

All the other colours – a colour is just photons of light with different energy – may not be visible, but they are still part of this electromagnetic spectrum. Phones and Wi-Fi work in the radio band of the spectrum. Like all electromagnetic radiation, if there's too much of it, it can be dangerous – but at low levels there's no risk from it. Electromagnetic radiation in the form of radio is quite different from ionizing radiation, which is the stuff that comes from radioactive materials. It includes X-rays and even more powerful gamma rays, but also particles like electrons and the nuclei of small atoms.

Ionizing radiation is a problem for us because it can penetrate the skin and damage the inner workings of the cells in our bodies. Medium-size doses can increase the risk of cancer, while large doses will produce radiation sickness, and, at extreme levels, death. So this is something to take seriously. But it has to be put into proportion. We're exposed to radiation all the time. We can't escape it. Even if you sat in a lead-lined box, breathing filtered air, you would still have exposure to radioactivity. That's because you are naturally radioactive. Some radioactivity is produced by the material in your body, and this will occasionally damage the odd cell – but the risk is vanishingly small.

Similarly, there's plenty of natural radioactivity in the world around us. Some places are significantly more radioactive than others. Where, for instance, there's a lot of granite rock there's usually increased radioactivity and a slightly higher risk. This can be particularly significant if a radioactive gas from the rocks called radon builds up in a house. In locations like Cornwall in the UK or Denver in the US where this risk is larger than usual, extra effort is put into ensuring that houses are properly ventilated so radon can't accumulate.

Experiment NOT to try – You might have to wait a while for this one

We have all had circumstances when a computer takes us by surprise. I don't mean when you do something silly, or the software has a bug, but you might be doing something perfectly normal, something you've done thousands of times before, and for no obvious reason the computer crashes. While it's most likely to be a bug, this can be caused by natural radiation.

What has happened is that a radioactive particle or gamma ray has bashed a few electrons out of one of the chips in the computer. This throws the memory or the processor off kilter and typically results in a crash. Unfortunately there's no way to predict when

---

this is going to happen. It's probably best to give this experiment a miss, as you might have to wait several years.

Fooled by a natural high

A lack of awareness of natural radiation levels sometimes leads to scare stories. When the Three Mile Island nuclear power station accident happened in the US in 1979, amateurs with Geiger counters (devices that measure the level of ionizing radiation) panicked when they discovered that radiation levels around the plant were 30 per cent higher than the national average. This sounds scary. There were headlines in the US claiming dire problems. But the detector operators had got things wrong. The measurements would have been the same if the power station hadn't been there – the Pennsylvania site just happens to have relatively high background levels.

You can't see which areas have more radiation than others from the sky, so what has this to do with your flight? Another place where radiation levels are higher than usual is on an aircraft. This has nothing to do with the plane itself – once more, it's purely natural radiation at play. Where increased levels on the ground are due to radioactive rocks, in the air, extra radiation comes from cosmic rays.

A cosmic collision

Cosmic rays are high-energy streams of particles that crash into the Earth's upper atmosphere from the depths of the universe. Some come from the Sun, many more from the depths of outer space where they may have travelled millions of years, across billions upon billions of miles from distant stars and stellar explosions. When these particles smash into the upper atmosphere, the energy from the collisions with air molecules produces other particles and high-energy light, which smash down towards Earth.

Some of these particles disappear very quickly. For instance, particles called muons pop into existence only to disappear again in a fraction of a second. They're travelling so fast, though, that Einstein's special relativity comes into play. As we saw with GPS, this says that for things that move very quickly, time slows down. Very few of the muons should make it to ground level, but at the speed they are moving, time slows down for them by a factor of five, giving many more of them a chance to make it through – a real demonstration of relativity in action. But the thing that causes more concern for air travellers is the high-energy light that's produced – X-rays and the even more powerful gamma rays.

In the average trip across the Atlantic, your exposure to this radiation is the equivalent of having a chest X-ray, amounting to around 100 times your typical dosage on the ground. Exposure to such radioactivity is measured in milliSieverts (mSv). The average background radiation you get in the UK or the US is around 2.5mSv a year, rising to around 7 or 8 in areas like Cornwall or Denver. If you fly ten hours every week, it adds around 4mSv to your exposure – so there's no more risk than moving from London to Cornwall.

However, exposure is cumulative, so if you're flying much more than ten hours a week – or combine a lot of flights with living in a high background area – it's worth being aware that you're exposing yourself to a degree of risk. If you want to cut down your background exposure, as well as moving away from high exposure areas, you could also cut shellfish out of your diet, as this can add around 0.5mSv a year to your exposure (because shellfish tend to filter natural radioactive materials out of the water).

The other thing to watch out for is the solar cycle. The amount of radiation the Sun pumps out varies on an eleven-year cycle. Depending on where it is in its cycle, the Sun can add between almost nothing and 1mSv an hour during a flight. There's a maximum in 2011, a minimum around 2016 and the next maximum in 2022. Radiation exposure on any individual flight isn't anything to worry about, any more than worrying about holidaying in Cornwall, but if you're flying more than once a week it's worth finding out a little more.

Cabin Life

Pressure on the blood supply

Back inside the cabin, there are a couple of other better publicized worries that passengers face – deep vein thrombosis (DVT) and jet lag. Millions of people suffer from DVT each year, though mostly without serious outcome (and mostly not on flights). It's caused by a small blood clot, typically forming in your legs on a plane, due to the constant pressure from the seat on a long-haul flight, made worse by the reduced air pressure.

If you're on a long-haul flight, a little exercise is an excellent way to reduce the risk. Try to move around the cabin at least once an hour. If it's not practical, flex and massage your legs (in fact, try to do this regularly too, whether or not you get your stroll). It also helps to drink plenty of water and stay away from dehydrating substances like alcohol and coffee. If you don't have a problem with aspirin (check with your doctor), a low dose, typically half a tablet for an adult, can help by thinning the blood a little. And you can also go for those unflattering support stockings, which can be worn unobtrusively under trousers.

Catching up with jet lag

Jet lag may be more of an irritation than a threat to life, but it's something most travellers who cross multiple time zones experience. These divisions of the Earth into separate time bands were introduced as the railways changed our approach to clocks. The use of time zones was an arbitrary decision. We could have had a fixed time for the whole planet, where 17.00 hours would be in the morning in some places, in the afternoon in others and elsewhere would be in the middle of the night. Times would be the same everywhere, but the time would no longer indicate a particular part of the day. This wouldn't help with jet lag, though, as our waking day would still be tied to daylight, and so different around the world.

Crossing the time zones

Before time zones were established, everyone operated on their own local, sun-based time. There was no coordination of time, which meant it could and did vary from city to city. Twelve noon in New York was different from twelve noon in Boston. London time differed from Birmingham time. But the coming of the railways made it essential to be able to exchange times between different stations along a line. You could hardly have a timetable where each arrival and departure reflected the different idea of when twelve noon was in that particular city. Time zones as we now know them were mostly introduced in the late 19th century. The current US time zones, for example, were formalized in 1883.

The main time zones divide the world into segments that share the same time of day. The USA, for example, has four mainland time zones (excluding Alaska), from Eastern Standard Time at five hours after Greenwich Mean Time (GMT) to Pacific Time, which is eight hours after GMT. (A few locations around the world confuse matters by using half-hour and 45-minute variants in their zones. So, for example, Venezuela is 4½ hours behind GMT, while Nepal is 5¾ hours ahead.) If the time zones were like evenly spaced segments of an orange, each should be separated by 15 degrees of longitude, but in practice they are separated by wiggly divisions that veer here and there as they pass from pole to pole.

Some borders between time zones are at sea – and one particular border, the date line that separates one day from the next, is intentionally jagged so that it avoids pretty well all land masses. But with four time zones dividing up the US, it's inevitable that there are places where just taking one step means that you've moved an hour into the future, or an hour into the past. Stand on the boundary between Georgia and Alabama, for example, and stepping across that border will flip you between Eastern Standard Time and Central Standard Time. This game isn't possible in China, which is big enough to cover five time zones but has instead adopted a single universal time across the country.

One trick that time zones make possible is to freeze time. Fly at the right speed and you can spend a whole day flipping between New Year's Eve and New Year's Day. That speed depends on where you fly. You would have to go fastest at the equator, where you would be covering 40,000 kilometres in 24 hours. That would mean a speed of 1,666 kilometres an hour (just over 1,000 miles per hour). Too much for a present-day plane, though well within Concorde's capabilities. But if you circle the globe near one of the poles there's no difficulty keeping up, as the distance gets smaller and smaller until at the pole itself you can traverse all 24 hours simply by turning around. If your plane managed to land on one of the poles you would, in principle, be in 24 different hours, all at the same time. (This can't work in practice, so both poles use GMT as their official time.)

What jet lag is (and isn't)

When it comes to jet lag, though, the individual time zones aren't a problem. All we're really concerned about is a combination of tiredness and confusion caused by doing daytime things at what would, for your body, usually be the middle of the night – the same kind of problem experienced by shift workers. Jet lag isn't a medical condition and isn't something that can be 'cured', any more than being tired can be cured. The impact can be minimized or you can fool your body into ignoring it, but you can't take a magic pill and make it go away. The best that can be done for jet lag is the equivalent of taking a stimulant to stop feeling tired.

Even though jet lag isn't an illness, it can have a serious impact on your ability to concentrate and make sensible decisions (again, just as fatigue can). A classic example of the danger of jet lag comes from the late John Foster Dulles, one-time Secretary of State of the USA. Dulles admitted in later life that he had precipitated the Suez crisis in 1956 by making a decision to cancel a loan to Egypt for the Aswan Dam immediately after his return to Washington from the Middle East. He didn't allow the time to recover from his journey and considered jet lag to have been the primary cause of his mistake.

If you have enough time, you can just sit jet lag out. It takes around a day to recover from each hour of time zone crossed, with the effect being less strong when travelling in a westerly direction. Adding to the day by catching up with the Sun seems less traumatic than chipping away at daylight. If you want to do more than just wait, though, there are some actions you can take to reduce the impact of jet lag, and it all starts with food and drink.

Taming jet lag

Ideally, begin your attack before flying (this advice isn't much use if you're reading this on board, but there's always next time). Get as much sleep as you can before flying, keep your food intake bland for 24 hours before, and drink sensible amounts of fluids. (You may have heard that water is the only fluid that's good to drink for hydration, because even squash or fruit juice don't give you the same effect as water. This is rubbish. But avoid diuretics – drinks that encourage your body to lose fluids – which include coffee, tea and particularly alcohol.)

When you get on board, set your watch immediately to your destination time and stick to that. Try as much as possible to fit the way you behave on board around the time you will arrive. So if you're due to land in the morning, try to get at least six hours sleep before getting there. If you're arriving in the evening, try to keep awake for the previous eight hours. Stick to this routine on arrival. If, by local time, this isn't a time you would normally have a nap, don't have one.

The main inconvenience on the plane is that you may well not be able to fit with the meal schedule – but stick as much as you can to the meal times that work for your destination. If that means going without food that's on offer, resist the temptation to tuck in. If you're in economy, you probably aren't missing much, and if you're in a more expensive seat, the cabin crew should be accommodating enough to bring you food when you want it, rather than when their schedule dictates.

To be honest, it's best not to eat too much on the plane, anyway. And ideally keep entirely off the booze – alcohol dehydrates your body and increases the impact of jet lag. Keep colas, coffee and tea to a minimum too, as caffeine is also a diuretic. If, when you arrive, it's daytime, take a little walk – exercise will help hold off the jet lag. Make sure you eat at local times, not by the schedule that seems right for your departure point.

Resorting to medication

There have been a number of studies into the benefit provided by the anti-jet lag drug melatonin, which you may have seen advertised on the internet. This pineal hormone is said to fool the brain into ignoring the impact of changing time zones. Research at the University of Heidelberg has shown that after a long flight there were changes in the body's levels of melatonin, and there's a strong link between melatonin levels and the ability to sleep. But there are real questions attached to the prescribing of melatonin as a 'cure' for jet lag. (Remember, jet lag isn't an illness – you can't cure it.) According to the UK medical journal, *The Lancet*:

> [Melatonin's] apparent usefulness in alleviating the effects of jet lag may be related to some ill-defined psychotropic activity, but such effects may

be undesirable, particularly if they modify day-time function [...] Melatonin is a possible inhibi-tor of sexual development in rats and may initiate gonadal regression in voles. It also has endocrine effects on man. The effects cannot be dismissed lightly ...

If the drug works at all, it's necessary to undertake a rigid timing programme of doses around your flight time. Get the schedule wrong and the drug will make jet lag worse instead of improving your response. More worry-ingly, however, as with any hormonal treatment, there's considerable concern about possible side-effects. When compared with a non-invasive approach like the sugges-tions about timing meals and sleep, the use of melatonin seems doubtful at best.

There's at least some evidence that melatonin has an effect, even if the possible side-effects are worse than the original problem. By comparison, some suggested jet lag cures are based entirely on fooling ourselves. There's a whole range of alternative therapies and exotic treat-ments available. Some involve a complex diet process with fasting and feasting for days in advance of the flight. Others involve shining bright lights on the traveller to confuse their natural rhythms. Perhaps the only ones worth considering are aromatherapy and homeopathy.

Aromatherapy and homeopathy don't do any-thing whatsoever to the body. (This isn't surprising. Homeopathic remedies, for example, will usually con-tain none of the active ingredient mentioned on the bot-tle.) But they are good at stimulating the placebo effect.

This happens when your brain believes that you're gaining a benefit, and releases natural chemicals into the body that can have a real effect. Using such placebo remedies can be seen as a complement to the simple sleep and water approach, and certainly won't do any harm. Anything that helps you get off to sleep in a natural way should be beneficial.

Is there a jet lag north/south divide?

Some exotic ideas have also been used to explain the less documented problem of north/south jet lag. Conventional jet lag arises on flights that involve crossing time zones, producing an extended or shortened day. But what of long-distance flights that don't result in a change of time, where the travel is all within a single time zone, say from Toronto to Lima, or Johannesburg to Helsinki? If time-shift were the sole cause of jet lag, there should be no effect produced by such long-distance flights from north to south or from south to north. But there is.

This has led some writers on air travel to suppose that bizarre forces are at work on the body. They point out that travelling on a north/south line cuts through the Earth's magnetic fields. They note that water (a major part of the body, after all) seems to act differently north and south of the equator, as shown by the way that it runs down the plughole clockwise in the north and anti-clockwise in the south. The conclusion drawn is that being in a different hemisphere may have a direct effect on our bodily fluids.

Unfortunately, these entertaining ideas have little basis in scientific reality. The bathwater's direction of spin would be caused by Coriolis force – as we've already seen, a side-effect of the Earth turning round (see page 37). This force is very weak and all the evidence is that the supposed difference in direction of water going down the plughole is a myth. The influence of Coriolis force on the body is incredibly slight, much less than any influence from the movement of the plane. As far as the magnetic field of the Earth goes, though there's good evidence that some birds can detect changes in magnetic fields and use these in navigation, the Earth's field isn't strong, and again there's no obvious mechanism for it to have an impact on our well-being.

Instead of being baffled by pseudo-science we just need to look back at what we've decided jet lag really is. It's exhaustion similar to that experienced by a shift worker, coupled with a disorientation from time change, the impact of dehydration and exposure to low air pressure. All of these factors except for the time-based disorientation are still present in a flight that stays in the same time zone. There's no need for any mystical force to provide a weaker form of jet lag in these north to south and south to north long-haul passengers. Nor is there a need for a different cure. The same things will help with this form of jet lag.

A moving experience

We've been concentrating on the inside of the plane. It's time to take another look out of the window. Anything nearby – a cloud, for instance – seems to be rushing past at great speed, while things in the distance hardly seem to be moving at all. What's going on here?

Experiment – Playing with parallax

Look out of your plane window and hold up a finger, in front of your face, a foot or so from your eyes. Shut your right eye and notice where your finger falls. Now switch eyes, closing your left eye and opening your right. Flick between the two eyes several times. You will see that the finger moves backwards and forwards with respect to the window and the view outside. Obviously your finger isn't actually moving back and forth, but changing your viewpoint adjusts its apparent position. Your eyes are differently positioned, so they see the finger a different way.

This effect is known as parallax, and it can produce even more confusing effects. If you can see clouds or landmarks below, fix your eyes on the middle distance. Watch for a while as the plane flies on. How does a more distant cloud or landmark move with respect to the point you're looking at? How about a closer point? The more distant point seems to move *forwards* compared to that mid-point as you fly along, while the closer point seems to go backwards.

Because our eyes see the world from two different points, something that's essential to get 3D depth of vision, switching between the eyes shows us two different viewpoints. The further away something is, the less it seems to move when we switch angle of view, so your finger seems to move against the background. Think of a line stretching from your left eye to a distant object. Now think how your line of sight would move when you switch to your right eye. The line sweeps out more distance near your eye than it does in the far distance. Similarly, the close-up clouds, sweeping out a much bigger angle in the same time, will seem to shoot past, while the more distant ones hardly move.

If you think of that same image of a line connecting you to a distant point, and pin the line down on the point in the middle distance you fixed on in the second part of the experiment, then as your view swings from left eye to right eye, the distant object will get ahead of your line. It will seem to move forwards compared to the middle distance, even though it's still going backwards with respect to the plane. Things that are closer than the middle distance are left behind by the line – they move backwards compared to your mid-point. In practice, those different objects outside aren't moving with respect to each other. It's just your changing viewpoint that makes it seem that way.

Relatively interesting

When something really does move with respect to something else, as you're moving with respect to all the stuff outside the window (apart from bits that are stuck to the plane, like the wings), the impact that movement has is called relativity. If you aren't flying with a budget airline, you might be served some refreshments around now, which gives us a good opportunity to experience relativity in the aircraft cabin.

Albert Einstein famously described one type of relativity. He said: 'When a man sits with a pretty girl for an hour, it seems like a minute. But let him sit on a hot stove for a minute – and it's longer than any hour. That's relativity.' If you're sitting next to someone on the flight you may have experienced either of these effects – time passing by quickly as you sip a drink and chat with a pleasant companion, or dragging out interminably if you're stuck next to a bore.

Although Einstein jokingly called this effect relativity, what he was really describing is subjective time – our experience of the passage of time, which can be quite different from the steady ticking of the clock that marks objective, scientific time. This is 'a watched pot never boils' time. However, the real physical version of relativity can also be observed from your plane seat in many different ways.

Galileo's big idea

Despite the way we think of Einstein the moment that relativity is mentioned, he didn't invent the concept. It was Galileo who came up with the modern basics of relativity in the 1630s, totally amazing those around him.

Experiment – Ice breaker

You'll need a drink with a piece of ice in it. Stand the drink on your table and watch the ice. If you have the chance, see what happens if the plane is in steady, stable flight. Also keep an eye on the ice if there's turbulence, or if the plane accelerates or turns.

When the plane is flying steadily, without any acceleration, the ice floats motionless, just as it would on the ground. If the plane is really steady, with the window blinds closed you wouldn't be able to tell you were moving at all. You couldn't do any basic physical

---

experiments inside the plane that would tell you that you were moving. Things behave entirely normally. This was Galileo's big idea. He used a steadily moving ship as his example, but it's the same concept.

When the plane encounters turbulence or accelerates (turning is just a form of acceleration), suddenly things don't behave normally. We can tell we're in a moving vehicle, whether from the way the ice acts or from the impact on our senses.

Galileo realized that all movement has to be relative to something else. If two objects are both moving at exactly the same speed in the same direction, each object sees the other one as stationary. If there's another plane in sight, and you're both travelling in the same direction at the same speed, the other plane will look to be standing still. But it's more than just a matter of looks – that plane really *is* standing still from your viewpoint. If they were close enough you could step from one to the other without feeling anything.

One of the reasons this was so important to Galileo was in his support for Copernicus' idea that the Earth moves around the Sun. Ever since the ancient Greeks it had been assumed that the Earth was motionless at the centre of the universe with the Sun (and everything else) moving around it. A common objection in Galileo's day to the idea that the Earth was moving was that anything not fixed down would fly off the Earth to be left in its wake. Yet this doesn't apply because from our viewpoint, the Earth isn't moving. We're all moving at the same speed as the Earth (unless, for example, we're up in a plane).

Experiment – Catching up

We're going to recreate an experiment Galileo under-took, amazing a group of individuals who didn't understand relativity. The party were out in a boat on Lake Piediluco, rowed at speed by six men. Galileo asked a friend, Stelluti, if he had anything heavy to hand. Stelluti produced a heavy key, obviously important and difficult to replace. Galileo astounded him by throwing the key as hard as he could straight into the air. Stelluti panicked and nearly jumped out of the boat, certain that the key would be left behind as the boat moved forward and drop into the water behind them.

Don't try to throw keys above your head on a crowded plane. Instead, use a scrumpled up ball of paper – and move to an open area away from passengers and crew. Throw the object up in the air as straight as you can. Galileo's friends would have assumed that the object would shoot backwards down the plane because the plane is being pushed forward and the ball isn't. This clearly doesn't happen.

Until Galileo's time, everyone assumed that things kept moving only if you pushed them (or if they were under the influence of gravity or levity). Clearly the ball is moving with you (as the key was moving with Galileo) until you throw it upwards. But then, it was assumed, it would stop being pushed along with you and so would be left behind. In practice, as you should have seen with your ball unless you failed to throw it straight up, there's no relative motion between you and the ball, and there's nothing acting on it to start it moving with respect to you.

Let's imagine you somehow got onto the wing of the aircraft with your ball of paper and again threw it straight up. What would happen this time? Here the ball *would* shoot backwards. But this is because the aircraft is moving quickly with respect to the air. From your viewpoint, sitting on the wing, the aircraft is stationary and the air is moving backwards at high speed. That backward-moving air will hit your ball of paper and the many billions of collisions with high-speed air molecules will propel the ball backwards. Galileo used a heavy key on the lake to avoid this problem.

In the jet stream

When the plane was low and you looked out over the ground, it was very clear that you were moving forwards with respect to the Earth. Or, as Galileo might have pointed out, you could equally say that you were sitting still and the Earth was moving backwards under you. (Apparently Einstein had a habit of asking, 'What time does the station arrive at this train?' The joke probably got a little irritating after a few repetitions.) But how fast are you going?

If you're in a 747, for example, the cruising speed is between 550 and 570 miles per hour (around 890 km/h). Yet if an observer down on the ground used a radar speed gun on you, and you were heading from the US to the UK, she could well decide that you were travelling faster than sound. You can – relatively speaking – break the sound barrier in a conventional plane without even noticing it.

The airspeed of an aircraft is its speed relative to the surrounding air, not relative to the ground. That cruising speed of 550mph is how fast you're going compared with the air around you. But say that air itself is moving in the same direction you are, at 200mph. Then your speed, as seen by the radar gun on the ground, would be 750mph. You would be travelling above sound's 740mph velocity.

Such wind speeds may seem extreme, but they're not uncommon in the powerful jet stream – a constant current of air that moves from west to east as a result of the rotation of the Earth. Jet streams form at a temperature inversion in the atmosphere, in a region where temperature increases if you go higher or lower – something that typically happens around the 30 to 40,000-foot cruising height of an airliner. This is the border between the lower atmosphere, the troposphere, where the greenhouse effect dominates, and the upper atmosphere, the stratosphere, where the direct rays of the Sun have the biggest impact. The jet streams don't occur everywhere around the globe, but form in long corridors and can flow as fast as 250mph. This can make eastbound flights at the right location significantly faster (and more fuel-efficient) than their westbound counterparts, and even enable a supersonic flight.

The special one

In 1905, Einstein took relativity even further than Galileo with one remarkable discovery. Let's go back to the jet stream. Remember the way the speed of the air added to that of the plane to produce a faster speed over the land. Similarly, Galileo's relativity tells us, if two planes fly towards each other, then we can add the two speeds together. Each might be flying at 550mph with respect to the ground below them, but from one plane, the other plane is travelling towards it at a massive 1,100mph. And, as we have seen, if two planes fly alongside each other at the same speed and in the same direction, they don't move with respect to each other.

At the time he came up with his discovery, Einstein wasn't thinking about planes (which had only just been invented, as this was a mere two years after the Wright Brothers' first flight in 1903), but about light. Imagine your plane suddenly becoming incredibly quick. You're flying along at 186,000 miles per second (300,000 kilometres per second). That's the same speed as light. Alongside you is a sunbeam heading in the same direction. According to Galileo, from your viewpoint that sunbeam would be stopped. But Einstein realized that this would be a huge problem.

The Scottish scientist James Clerk Maxwell had shown towards the end of the 19th century that light *has* to travel at a specific speed. Its speed varies in different substances (it's faster in a vacuum than it is in glass, for instance), but in any particular medium – air, say – it has no choice about how fast it goes. Light is the interaction of electricity and magnetism. If you move an electric current, you generate magnetism. Move a magnet and you make electricity. Move your electricity at just the right speed and electricity makes magnetism which makes electricity, which makes magnetism and so on, hauling itself up by its own bootstraps. But this works only at the specific speed of light. Light can't exist at any other speed.

So Einstein realized that if light changed speed as we move with respect to it, the light would disappear. As soon as your plane started moving, all the light outside it would fall apart. Anyone who wasn't perfectly still (with respect to what?) wouldn't be able to see, as all the light around them would disintegrate. That's ridiculous – it just doesn't happen. So Einstein had the wild idea that light always travels at the same speed, however you move with respect to it. When you're flying in the direction of sunlight at 550 miles per hour, the light doesn't get 550mph faster, it comes at you at exactly the same speed as it comes towards a plane that's parked on the ground.

When Einstein plugged this idea into Newton's equations that explain motion, he got a shock. To be able to fix the speed of light he had to loosen up a number of other things that everyone assumed were unchanging. If an object moved, then his new theory said that it became heavier, it shrank in size, and time slowed down for it. This effect is very small unless you travel near the speed of light, but the consequences are big – as we have seen, GPS satellites wouldn't work if they didn't correct for it.

Anti-ageing flights

Einstein called this concept 'special relativity' because he had looked at the special case of relativity when we're dealing with things moving steadily without accelerating. The strangest implication of his theory was what has been called the twins paradox. If we take a pair of twins, send one off in a spacecraft at near the speed of light and leave the other on the Earth, time will pass more slowly for the one on the spaceship. When she gets back to Earth she will be younger than her twin. This isn't a subjective effect, like the story with the hot stove – this is real slowing down of time. The experiment has been done not with twin humans but with twin atomic clocks, and the one that travelled did record time getting slower.

As this applies to any speed of travel, when you get back from your plane journey, you too will be very slightly younger than you would be if you had stayed at home. If you have a twin, you will now be a tiny smidgen younger than your sibling. The effect is absolutely minuscule at a plane's speed. To have a noticeable effect, you would need to travel at a reasonable percentage of the speed of light. But even at humble aircraft speeds, after 40 years of crossing the Atlantic on a weekly basis you would be one thousandth of a second younger. Not, I admit, something for the anti-ageing cream manufacturers to worry too much about.

A nice cup of tea

We started our exploration of relativity with a drink, but another drink you might be served on board could be a little disappointing – and not just because you might have to drink it out of a plastic cup. That's a nice cup of tea. Tea enthusiasts like their tea made with boiling water – which means getting the water up to 100°C. That's never going to happen on a plane. Not because the cabin crew can't be bothered to do it properly, but because it's impossible get water up to 100°C on board the aircraft.

This is an effect of the air pressure. Just think what's happening when a liquid like water boils. It means the molecules in the liquid are moving fast enough to be able to escape into the air. If there wasn't any air, they could be moving at any speed and escape – and if you take water into the vacuum of space it will boil, despite the intense cold. Usually, the air pressure stops most of the water molecules escaping. Imagine the air being like a bombardment of billions of balls, all battering down to keep the water in place. Reduce the air pressure and you cut down on the number of balls hitting the surface, so the water molecules don't have to go so fast to force their way through and escape.

At the pressure of an aircraft cabin – the equivalent of being up to 8,000 feet above sea level – water boils at around 90°C, and that's as hot as your tea is going to get. No matter how much heat the cabin crew pump into the boiler, it won't shift above 90°C until all the water has boiled off. This is because any extra energy you push into water at its boiling point goes into breaking down the bonds that hold water together as a liquid, forcing more of it to turn to gas, rather than changing the temperature of the water. It's only once all the bonds are broken that the temperature will begin to rise again.

Imagine if you had one of the water boilers from the galley on the wing of the aircraft. A cup of tea made with this water would produce a very unappetising drink. With the reduced air pressure at around 40,000 feet, water boils at just 53°C.

So if you only like tea that has been made with 'proper' 100°C boiling water, it's probably best to stick to coffee on your plane journey.

Hearing food

If the tea disappoints, the chances are that the food will as well. This probably isn't too much of a surprise. Unless you're travelling in Business or First Class, airline food doesn't have a great reputation. Yet this isn't necessarily the fault of the airline. There are physical effects that can reduce the tastiness of the food. There could be some impact from low cabin pressure, the very dry air on board and the way meals are reheated, but one surprising cause is the level of background noise.

Although we get used to it quite quickly and ignore it, there's a constant noise in flight, both from the engines and the air systems (not to mention fellow passengers). If you're on a plane as you read this, you'll suddenly begin to notice that noise again. It's always there. Research published in 2010 looked at the way noise influences our experience of food. Test subjects were blindfolded and asked to rate various foods as they were exposed to differing levels of background noise. When the noise was louder they tended to rate food as less sweet or salty, but more crunchy. However, the research featured a relatively small sample of 48 people, so there's more work to be done before making a definitive decision as to whether airline catering would be improved by issuing earplugs.

Technology in Flight

Following your course on the map

After a while, stunning though it may be when you really look at it, the view from your window becomes a little samey. It's time to pay a bit more attention to your inflight entertainment. The chances are that somewhere there will be a display of your flight's progress with a little electronic map showing the course the plane is taking. This route will often appear bizarrely indirect. It would seem sensible to follow a straight line when fuel is so expensive. It's not as if there's much in the way, up in the air. Yet airlines insist on flying in curves. Three reasons conspire to take you off that simple straight line.

The most fundamental of these is that the shortest distance between two points on the Earth usually won't appear as a straight line on a map. The Earth is roughly a sphere (it bulges out at the middle because it's spinning round, but we can ignore that for our purposes), and the shortest distance between two points on a sphere is part of a great circle. This is a circle that cuts the sphere it's drawn on in half – it has the same radius and circumference as the Earth itself. So a great circle is like an equator, but one that can be drawn between any two points on the earth, leaving a hemisphere on each side of it.

It might seem reasonable that two points on very different latitudes (your latitude is your distance from the equator towards one of the poles) would be best joined with great circles, but what about two points at a northerly latitude, both on the same line of latitude – say, circling the North Pole a few hundred miles from the pole? Wouldn't the distance be shorter running around the latitude line, following a smaller circle? The answer is no, because a 'small circle' like the latitude line has more curvature than a great circle, so increases the distance flown. In effect, to fly that route your line has to bulge out more.

To display your course on a screen (or in a flight magazine) you have to get your great circle off a sphere onto a flat map. The process, called projection, inevitably distorts shapes. To get from a globe to a flat map, each point on the map has to be transferred to a point on the flat sheet. The traditional way to do this is called the Mercator projection after the Belgian geographer Gerardus Mercator, who came up with it in the 16th century. His approach involves unwrapping the surface of the Earth onto an imaginary cylinder that's fitted around the equator. Like all projections it has significant limitations. Some projections are better at showing areas, others better at maintaining shapes as they appear on the surface of the globe.

Projecting the world

When making a projection like the Mercator, only a journey along the equator will appear to be a straight line – any other great circle will appear as if it's curving out away from the shortest distance 'straight line'. So, for instance, a flight from London to New York will seem to curve up towards the North Pole on the flight display, rather than take what appears to be the shortest route – yet in doing so it's minimizing the distance.

However, the simple matter of shortest distance isn't the only factor involved in choosing a flight path. Sometimes routes have to move away from a great circle for political or practical reasons. It isn't always advisable to fly through a particular country's airspace, or it may be that air traffic control takes flights through a particular corridor in the air to make them easier to manage and so restricts the pilot's ability to minimize distance.

We've already met the other factor that's likely to change the direction taken – the jet stream. As we have seen, when a plane is flying west to east it's likely to encounter high-speed winds that can significantly increase speed over the ground. Planes will often take a different route going west to east than they would in the other direction in order to stay in the jet stream and get the maximum boost to reduce journey time and fuel use. It's a bit like using a motorway – the extra speed more than compensates for any increase in distance.

At the bleeding edge of technology

That on-screen map is entertaining to check up on briefly, but it isn't going to keep you amused the whole flight. These days, most planes have a whole range of inflight entertainment. Oddly in such a high-tech environment as a plane, you may well find that the technology lags well behind anything you experience in your home, or that you can do on a smartphone. This is because planes are such huge construction projects that by the time they are built, much of the leading-edge technology inside is already out of date.

When Concorde was still flying it often amazed people just how primitive the flight deck was. In an age of digital cockpit displays, Concorde's dials and readouts were all mechanical – because it was designed in the early 1960s, before the IT revolution. (For a more extreme version of technology lag, you only have to look at the computing power on the Apollo Moon shot. Predating PCs, the onboard computer technology had less power than the most basic modern phone, let alone an iPhone or other state-of-the-art device. The Apollo Guidance Computer had the equivalent of 4 kilobytes of memory.)

Even a less revolutionary plane than Concorde is much slower in the planning and building than the entertainment technology we have in the home, which means that the inflight entertainment on a brand-new aircraft can still be surprisingly last-year. Think how much phones have changed in the last few years – yet during that time, most of the technology in the latest aircraft will already have been fixed.

Keeping the screen flat

For a while now, one particular technology has transformed inflight entertainment – the LCD screen.

Screens at the seat were impossible to implement until this technology came along. LCD has become so pervasive – it's hard to buy a TV or a computer screen that isn't LCD now – that it's hard to remember that this technology has only been available for mass use in screens since the late 1990s.

29. Seatback inflight entertainment.

Up until then, our TVs and screens would be CRTs or cathode ray tubes. This was a Victorian technology. A 'gun' well behind the tube shoots a stream of electrons out into the vacuum that fills the tube. This electron beam passes between a number of electromagnets. These focus it and move it side to side and up and down. Using the beam as a kind of electronic pen, the tube draws an image on a series of 'phosphors' on the screen. These are tiny dots of a material that glows briefly when it's hit by electrons. The beam sweeps over the screen so fast that a picture is built up from these dots.

The trouble with CRTs is that the gun has to be well back from the screen to give those magnets a chance to move the electron beam around. This is why old TV sets were so bulky – and why you could never have fitted a CRT screen into a seatback. That all changed with LCD. The initials 'LCD' stand for liquid crystal display, and it's a particularly cunning piece of technology, depending on a peculiar behaviour of light that was first noticed over 300 years ago.

Bartholin's crystal wonder

It was back in 1669 that the Scandinavian experimenter Erasmus Bartholin noticed that a kind of transparent mineral called Iceland spar split the image you saw when you looked through it into two. If you put a piece of this calcite crystal (a crystalline form of calcium carbonate, the main component of chalk, limestone and marble) on a piece of writing, you would see two copies, split apart from each other through the block. Bartholin thought that this was because there were two different kinds of light – and in a way he was right. What he didn't realize (unlike Edwin Land who invented Polaroid sunglasses) was that each photon of light has a direction associated with it, at right angles to the direction it's travelling in, called its polarization.

Ordinary light, coming from the Sun, say, consists of photons polarized in any old direction, an overall mess of all possibilities. But when light reflects off a surface, much of it is polarized in one specific direction. Land realized that if he made sunglasses with a filter in them that cut out light in that particular direction, they would reduce the amount of glare from reflections off the road and windscreens. Land became a millionaire – but his use of polarization in filters would be put in the shade by the technology that makes those liquid displays possible.

Experiment – Polarizing with a twist

For this experiment you need a pair of old Polaroid sunglasses that you don't mind destroying. Pop the lenses out of the frames. Hold one of the lenses up and look out of your window. If the lens is tinted, you will see a darkened image of what you saw before. Now put the two lenses together, both the same way round. Again, look through the window. Once more you should see the usual view, but now tinted even more because you're looking through two lenses.

Holding one lens still, rotate the other lens gradually until it's at 90 degrees to the first one. As you turn the lens, the view should darken until you can see nothing at all. There's just blackness. Rotate the lens further, and your view will start to reappear.

To get an idea of what's happening with that pair of Polaroid lenses, imagine each one as having lots of slots in it, going (say) left to right, which photons that are the right shape (elongated left to right) can fit through, just like shapes going through a child's shape sorter. The photons hitting the lens that are elongated up and down won't go through – so they're cut out.

When you turn one of the lenses through 90 degrees, anything that can get through the first set of slots can't get through the second set. The lenses are rejecting both light that's polarized horizontally *and* light that's polarized vertically. The result is that all the light is cut out, and the view goes dark.

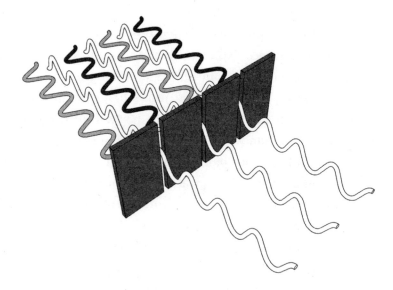

30. A polarizing filter splits off photons with a single orientation.

Giving light the liquid crystal twist

There's one more bit of technology we need, though, before we can make a liquid crystal display. That's the liquid crystal itself. This is a weird substance that, as the name suggests, behaves partly like a liquid and partly like a solid crystal. Some liquid crystals have the interesting property that when you put an electrical current across them, they rotate the polarization of light that passes through them by 90 degrees. So at last we're ready to build a liquid crystal display.

Here's the twist (literally). Imagine sticking a liquid crystal between the two Polaroid lenses when the lenses are turned at 90 degrees to each other. At this point there's no light coming through because the filters stop all directions of polarization. But put some electricity through the liquid crystal and it starts twisting photons through 90 degrees, like turning a screwdriver blade. If light is coming from the back, the 'sideways' photons will fit through the back slots, will be twisted through 90 degrees and then will fit through the front slots. Light will pass through your pair of lenses again.

In essence, that's how an LCD screen works. There's a flat, glowing panel at the back to provide plenty of photons. The light from it passes through a first polarizing filter, then a liquid crystal, then a second filter at 90 degrees. Result: no electricity = a dark screen; electricity on = a bright screen. Of course, it's more complex than this. The liquid crystal doesn't work on the whole screen at the same time. Instead, the screen is split up into millions of little segments that make up the pixels, the dots that form the picture. And each dot is split into three parts, one each for red, blue and green, to allow for a colour picture to be built up. But that's the basic workings in that seatback display, your phone, your computer or your home TV.

Taking your hi-tech with you

It's quite possible that you will have a more sophisticated piece of technology in your pocket than any seatback video – a smartphone. These remarkable pieces of kit cram a computer, keyboard or touch screen, often GPS, Bluetooth, Wi-Fi, a compass and more into a tiny package. Oh, and they're phones too. Many smartphones have a 'flight mode' that switches off the broadcast technology, leaving it as a pocket computer.

There's considerable confusion over just what electronics you can and can't use on board – and for that matter, whether the ban is really justified. In part, this confusion results from a lack of international standards. The airlines are largely left to set their own policies. While some flyers find it outrageous that they are limited in their use of electronics, it's probably best to be inclined to the cautious. The concerns about using electronics on board have no absolute proof attached, but it's hard to imagine that whatever you need to do with your phone or laptop during take-off (say) is so crucial that you want to put the plane at risk.

The problem is that modern planes are highly dependent on sensitive electronics, much more so than aircraft of 40 years ago. In principle, the radio signal from a phone, or the radio interference caused by other electronic devices like laptops, could interfere with these electronics, though like the ban on mobile phones in filling stations, there's very little evidence of danger.

All electronic devices may well remain banned from use in the particularly sensitive periods of take-off and landing. Most airlines allow laptops, phones in flight mode, ebook readers and the like at other times in flight, but there's something to be said for using the journey (during which your brain won't be functioning well anyway due to low air pressure) for something that keeps you away from the gadgetry.

The view from the flight deck
There was a time when one of the standard ways to pass the time on a journey (especially if you had children

with you) would be a trip to the flight deck or cockpit. This is likely to be off-limits now, but some aspects of inflight science were best viewed from there. From the flight deck, the view is so much better. Apart from having a panoramic outlook, rather than an elongated porthole, the flight deck windows don't have the inner plastic layer that protects the window glass from damage by passengers. It's this plastic that renders the view fuzzy and unimpressive from your seat – the view from the cockpit is startlingly clear and often stunning.

31. An aircraft flight deck.

There's plenty of technology up front, now all electronic, providing the controls and instruments, though relatively little of this intrudes into interesting aspects of science. However, it's hard not to feel a certain fascination with one piece of technology that you will find on every airliner's control panel – an autopilot (this is legally required to avoid crew fatigue on long flights).

A modern autopilot is primarily a dedicated computer, programmed to take readouts from instruments and guide the aircraft along a pre-programmed course which can involve turns, changes of altitude and more. Almost all of a typical flight can be undertaken on autopilot. There's an old joke among airline pilots that flight crew will soon be replaced by one pilot and a dog. The pilot is there to take over the controls if anything goes wrong, and the dog is there to bite the pilot if he tries to fly the plane when things are normal.

A lot of position information is now provided by GPS satellites, but the autopilot also takes information from an inertial guidance system or inertial navigation system, and here science comes to the fore.

Following the guidance of inertia

The key word in this system, which allows the plane to detect where it's going without reference to external sources like GPS, is 'inertia'. It's an aspect of the physics of motion that often causes confusion. Imagine you've got a heavy object resting on an ice rink. An elephant, say. You're wearing special boots that give you plenty of grip on the ice. Although the ice takes away most of the friction that normally stops you moving an elephant, it still takes quite an effort to get it to move. Once it's moving, you can't just stop it with the touch of a finger – something seems to keep the elephant going, and this is called inertia.

The interesting thing about inertia is that it depends on an object's mass, rather than its weight. We tend to use 'mass' and 'weight' interchangeably, but they aren't the same thing. An object's mass is an inherent property of that object, something that will be the same whether you're standing on the surface of the Earth, standing on the surface of the Moon, or floating in space. Mass describes how much force it takes to give that object a certain amount of acceleration. The bigger the mass, the more force it takes.

Weight, on the other hand, reflects the effect of gravity. So your weight will be six times greater on the Earth than it is on the Moon, and could be zero in space. As it happens, an object's mass on Earth is the same as its weight. This isn't a coincidence – it just reflects the way the units of mass were defined, so they come out the same on Earth, as that's where we tend to weigh things.

Because inertia is a product of an object's mass, it remains the same in space as it is on the surface of the Earth. It's just as difficult to stop (say) a truck moving along at 50 miles per hour in space as it is if it's rolling along without power at 50mph on Earth. In fact, it's harder in a way in space, because there's nothing for you to push against. Without that, no matter how strong you are, you have no way of stopping the truck. In space, even a superhero couldn't stop a meteor or an asteroid with nothing to push against.

Another reason inertia is confusing is that it isn't a force in its own right. Newton's second law (see page 39) tells us that to accelerate (or decelerate) a certain mass, you need to apply a force that is the object's mass times the acceleration you want. It's the need to apply that force that provides the inertia – there isn't a second force magically coming out of nowhere.

Tracking your way through the air

So how does inertia enable a navigation system to track a plane? An INS (Inertial Navigation System) starts by establishing a current position and speed, typically from GPS. It then measures the rate at which the plane is accelerating and decelerating, plus any angular acceleration (turning acceleration), to be able to keep tracking the location of the plane from its initial point. The advantage of doing this is that these measurements can be made even if radio communication is unavailable, leaving the plane isolated from GPS and navigation beacons. The INS will continue to plot the location with considerable accuracy despite any communication problems.

Discovering the values for these two kinds of acceleration requires different kinds of instrument. The acceleration in a straight line uses an accelerometer. These have become familiar as part of the sensing equipment of sophisticated mobile phones and games controllers. Whenever a Wii console's control senses the movement of the controller, or an iPhone registers the way it's being tilted and twisted, an accelerometer is in action.

The accelerometer relies on exactly the same principle as the one that inspired Einstein to come up with his second great theory, the general theory of relativity. We've already met the *special* theory, which is the one that says that as something gets faster, it becomes heavier, it shrinks, and time slows down for it. General relativity is the theory that explains gravity, and speeds up the clock on GPS satellites (because the stronger the gravity, the more time slows down, and satellites experience a weaker pull of gravity than we do on the surface of the Earth).

Einstein's accelerating revelation

The fundamental principle of general relativity came to Einstein while he was day-dreaming in his job at the Patent Office in Bern. It suddenly struck him that a person in free-fall doesn't feel their own weight. You may have seen this happening in a video of a 'vomit comet' – an aircraft that intentionally plummets to leave the inhabitants briefly in zero g. For a longer example, just watch a video of someone in orbit – for instance in the International Space Station. They float weightless in space.

You might think this isn't a great example, because the space station isn't falling – but in fact it is. Something in orbit around the Earth is always falling, it just happens to miss the Earth. The gravitational pull of the Earth is less at 300 to 400 kilometres up where the space station orbits, but it's certainly not zero. (If it were, the Moon, which is much further away, wouldn't stay in orbit.) However, an orbiting satellite like the space station does a clever balancing act. It's both falling towards the Earth in free-fall, and moving sideways (at a tangent to the Earth) fast enough that it always misses the Earth. If it travelled sideways any faster it would escape and fly away. If it went slower it would plunge to the ground. But at just the right speed – orbital velocity – it stays in its constant rotation around the planet.

Experiment – Orbital simulator

For this experiment you need a pair of earphones. (If you haven't got these, a smallish object with a bit of weight, attached to a piece of string, will do.) Hold the wire so the earphones are about two handspans from your hand. Spin the wire so the earphones rotate in a circle.

This is what happens to a space station, only we're making the gravitational pull visible in the form of the wire. The earphones (space station) want to fly off in a straight line at 90 degrees to the wire. But the force towards the centre down the wires (gravity) pulls them away from their straight line and keeps them in orbit. Note that there's nothing pulling them straight outwards (so-called centrifugal force) – the force, called centripetal force, is *inwards* to stop them flying off in a straight line.

What Einstein realized, thinking about gravity as he sat in his office chair, was that acceleration – getting faster and faster as you fall – and the effect of being under the influence of gravity are absolutely equivalent. This 'equivalence principle' works both ways. Being under the influence of gravity has the same effect as accelerating, and accelerating has the same effect as gravity. The most important thing Einstein would deduce as far as general relativity is concerned is that this means that gravity will bend a beam of light. Think of shining a light beam across an accelerating spaceship from the outside.

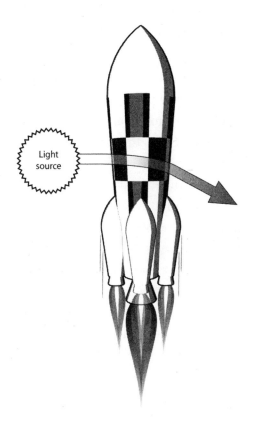

32. From within an accelerating spaceship, light bends.

As the beam crosses the interior of the spaceship, the ship's walls are being accelerated upwards, so the beam will strike the opposite side of the ship slightly lower than it would if the ship wasn't moving. As seen from inside the spaceship, the light beam will bend. But if acceleration does this, Einstein argued, so should gravity.

The only way Einstein could think to explain this was if gravity was twisting space itself – in fact, he deduced that gravity *is* a warp in space, produced by objects with mass, and this was what produced the effects of gravity. For an accelerometer, the equivalence principle means that if you measure a change in the pull of gravity, then that change is likely to be caused by an acceleration. (It's just possible that another body the size of the Earth has suddenly materialized in the sky and that has changed the pull of gravity, but it's more likely to be acceleration.)

At its simplest you could imagine an accelerometer as like a weight suspended from a spring that you hold in your hand. If you suddenly pull up or push your hand downwards, the spring extends a little more or a little less, making the weight seem to change. This change in the spring measures the acceleration. In practice, accelerometers are likely to use a more sophisticated way to measure the effect of gravity than a spring, employing a device that changes its electrical properties if it's stretched or twisted. This will act like the spring, but won't flop around as the device moves, and will produce a changing electrical current that can be used to measure the acceleration.

The feeble force

Gravity is just one of the four forces that are responsible for holding the universe together. Two of these forces, the weak and strong nuclear forces, work only at the scale of the insides of an atom, responsible respectively for the mechanics of nuclear fission and holding the nucleus of the atom together. The remaining force apart from gravity we've already met. It's the electromagnetic force, which we saw attracting bits of paper to a statically-charged pen. It also causes the attraction of magnets and all the familiar interaction between physical objects, such as when you sit in your seat and it holds you up.

We're used to thinking of gravity as being a powerful force. After all, it keeps the Earth in place in its orbit around the Sun, and everything firmly fixed on the ground. But in fact gravity is very weak. Just think – with nothing more than the muscles in your legs you can jump up and overcome the pull of gravity. Even more impressive was when you lifted a piece of paper with the electromagnetic force of the static charge on a plastic pen or comb. You had the whole, massive Earth pulling the paper down and just that tiny amount of electromagnetism pulling up – but still gravity lost.

Gyroscopic gyrations

So the accelerometer measures the rate at which speed changes in a straight line. When it comes to measuring the amount of acceleration in twists and turns, a gyroscope is often used. These fascinating devices, often featured in children's toys, usually consist of a fast-spinning disc mounted in such a way that it can twist and turn within its frame. The very simplest form of gyroscope is a spinning top. When the disc in the gyroscope (or the body of the top) is spinning quickly it has a considerable amount of angular momentum – the oomph that keeps it rotating – and this is something that is conserved. You have to do work to make it disappear.

We have already seen this conservation coming into play when water travels around a bend in a river. Conservation of angular momentum is also why a skater spins faster when they bring their arms in; the angular momentum depends on both the mass and how far that mass is from the centre, so by bringing the mass nearer the centre, the skater has to spin quicker.

In the case of the gyroscope, when the disc has a large amount of angular momentum because of spinning extremely quickly, it resists any attempt to change its orientation, because that would change the angular momentum. So when the disc is mounted in gimbals that allow it to swing freely in three dimensions, as you twist and turn the object the gimbals are attached to – an aircraft in our case – the disc will keep in its original orientation as the plane moves around it. By measuring how the orientation of the disc changes, you can monitor the twisting acceleration and keep track of the plane's flight path.

Distant Views and Back to Earth

Viewing the distant mountain peaks

Whether the autopilot is in charge or the human beings are at the controls, the views from the front of the plane can be stunning – and something that can be fully appreciated only from the flight deck is the full glory of mountains seen from the air. It's strange to think that until the 19th century the whole idea of 'natural beauty' really didn't exist. Wild countryside was seen as just that – wild and in need of taming. There was very little sense that there were views out there worth admiring. Yet now there are few things more heart-stirring than the sight of a mountain range glinting in the sunlight – and you can truly appreciate their scale only from the air.

33. Snow-covered mountain peaks (Himalayas, Tibet).

Just what a mountain is (as opposed to a hill) is totally arbitrary. There's no physical distinction between the two, leading to occasional fierce disputes over whether a local beauty spot qualifies as a mountain. The closest there comes to an international definition of a mountain provides a range of values that will make the distinction, depending on height, height above local 'ground level', and angle of slope, recognizing the feeling that mountains should be steeper than hills.

According to this approach, to be a mountain, an outcrop has to be a minimum of 300 metres above the local elevation (above what is regarded as 'ground level' in that area), or at least 1,000 metres above sea level with a slope of greater than 5 degrees, or at least 1,500 metres with a slope of greater than 2 degrees ... or anything more than 2,500 metres above sea level, no matter what the slope or where local ground level is. In US usage, a mountain has to be 1,000 feet above the surrounding land (taking the minimum to 304.8 metres), while in the UK, local height is ignored, and anything with a height of 2,000 feet (609.6 metres) above sea level gets mountain status. This demonstrates how much the definition of mountains is arbitrary – in Britain, where the tallest mountain, Ben Nevis, is only 1,344 metres high (4,409 feet), you need to be fairly generous with your definitions.

As mechanisms for establishing height become more accurate, we still get hills turning into mountains (and vice versa). In 2008, for instance, echoing the plot of the movie *The Englishman Who Went Up a Hill But Came Down a Mountain*, the Welsh hill Mynydd Graig Goch, thought to be 1,998 feet high, was discovered using GPS readings to actually be 2,000 feet and 6 inches (609.7 metres), scraping through to mountain status.

As old as the hills

Mountains and hills can be formed by volcanic action, when an eruption forces molten lava out from under the ground, piling up to form a distinctive cone-shaped structure (Japan's Mount Fuji is a classic example), but many are the result of the collision of tectonic plates. The Earth's surface sits on a number of vast plates of rock that float on the molten material below. These plates very gradually move, and where two of them head towards each other, the result is that the surface gradually crumples and is forced upwards to form a mountain range.

As you fly over mountains on the scale of the Alps or larger you will see snow-capped peaks all year round, even in the height of summer. It might seem rather strange, because the vast majority of the Earth's heat comes from the Sun – and the higher up a mountain you are, the closer you are to the Sun, so you might expect it to get warmer. There's certainly no doubt that the Sun's rays are strong on top of a mountain, requiring strong sunblock to avoid burning even if it's icy cold. So why is it so chilly up there? It's down to our old friend, the greenhouse effect.

It's cold on them thar hills

We've already seen how the greenhouse effect works. It acts like a blanket, keeping heat in that's trying to escape from the planet's surface. Not surprisingly, the thicker the atmosphere, the stronger the effect. By the time you get to 12,000 feet – around the height of Alpine mountains like the Jungfrau – the air is considerably thinner than at sea level. It's noticeably more difficult to breathe. That blanket of air that keeps the Earth warm is thinning out as you get higher, and this is why you will see those snow-covered mountain tops.

Roughly speaking, the temperature drops between 5 and 6°C for each 1,000 metres you climb. So let's say it's a pleasant 20°C down at sea level. You only have to get up to around 3,600 metres or 12,000 feet to reach 0°C. It should be no surprise that clouds often have ice crystals in them. If you're flying at 30,000 feet – around 9,000 metres – over an area with those same temperatures on the ground, it will be around –30°C outside the plane. This drop in temperature often ceases around this height, where you get the inversion in temperatures that holds the jet stream in place – but as far as mountains go, the rule is pretty constant.

The icing on the mountain

Where you get areas on mountains where snow can collect, pressure and repeated freezing and melting can convert the snow build-up into a glacier. This is a large volume of ice that typically moves slowly down the slope due to its sheer weight. Glaciers often stand out from the surrounding snow by having a blue tint. This isn't because the frozen water in them is any different from the surrounding snow, but simply happens because the glacier ice is more transparent than the snow, so more light can get through it. Just like water, ice absorbs rather more red light than the other colours, leaving any light that gets through it with a blue tinge.

At the end of the glacier the ice may drop off into the sea, calving icebergs, or melt, powering mountain streams. This melt water is essential for the fresh water supplies of some Asian countries – glaciers hold the majority of the fresh water that isn't in vapour form on the planet.

We tend to think of mountains as being ancient, long-enduring parts of the Earth, but some are relatively recent intruders. If we think in human timescales of a 70-year life, or the roughly 100,000-year existence of *Homo sapiens*, then, yes, mountains are ancient – but compared to the 4.5-billion-year age of the planet, some mountain ranges are mere youngsters, still settling in.

Take the Himalayas, our highest above-water mountain range. They have been in existence for just 1 per cent of the Earth's lifetime. Around 50 million years ago the tectonic plate that houses India ground its way into the main Asian plate. As the two pushed into each other, a new mountain range was thrust up into the air until it towered more than two miles above sea level. This range was high enough to interfere with the jet stream, making a massive change to the climate.

The intrusion of the mountains increased rainfall in the area, and these monsoon rains reacted with the carbon dioxide in the atmosphere. The acid formed by carbon dioxide and water dissolved the surface of the rock below, forming stable carbonate compounds that locked in the carbon dioxide, preventing it from returning to the atmosphere. In a reverse of our current problem of global warming, the temperatures began to drop. The formation of the Himalayas triggered a series of ice ages that transformed the world.

Around the bend with a siphon

If you had been to the flight deck, or have just taken a quick walk around to stretch your legs, you might want to call in at the toilet on the way back to your seat. Airline toilets are usually compact and significantly different in technology to the typical toilet you find in the home. All toilets are faced with the challenge of shifting a mix of liquid and solid matter away down a tube. In the home, this is most likely to be done with the aid of a siphon.

Siphons demonstrate some basic physics of air pressure that can appear almost magical.

Experiment – Siphon surprise

This is one you should leave until you get home. You'll need two glasses and a bendy straw. Fill one glass nearly to the top with water. Now suck up water into the straw until the straw is full of liquid. While keeping the suction going (this can be a bit fiddly), put your finger over the top of the straw. (Or, if it's easier, you can put your finger over the bottom of the straw, fill it with water, then put your finger over the top. Finally release your finger from the bottom.) You should now be able to lift the straw up and the water will stay in it despite the bottom being open, until you take your finger off the top, when it will drop back into the glass.

> Now hold the full glass in the air and put the shorter part of the bendy straw over the edge of the full glass and into the water, letting the longer end of the straw trail down over the second glass. Suck water through the straw until it's all the way through, then take your mouth away. The water will continue running through the straw from the full glass until the straw is no longer dipping into the water. You have constructed a siphon.

There are two interesting things happening when you do the straw experiment. The first is the way the water stays in the straw while your finger is over the top, despite the straw being open at the bottom. The other is the way that the water flows uphill into the bendy straw, out of the first glass and into the second without a pump needed to force it on its way once you've got it started.

The first part of the experiment shows how powerful atmospheric pressure can be. We don't usually notice the pressure of the air around us because it's always there, but it can do remarkable things. As usual, when thinking about movement, we have to follow Newton and ask what forces are acting. In the straw with the finger on the top, if we ignore friction, which is pretty limited here, there are only two significant sources of force – air pressure and gravity. Gravity is pulling the water downwards – that's nice and simple. There's pretty well no air pressure on the top of the straw because your finger is stopping the air getting to it. But at the bottom it's open to the air, so there's the full air pressure in action.

We can do some back-of-an-envelope calculations to see why atmospheric pressure wins. A straw full of water typically holds around 3.3 millilitres – conveniently, that weighs 3.3 grams, which is 0.0033 kilograms. Force, as Newton's second law showed us, is just mass times acceleration. So here we have the mass of water acted on by the acceleration due to gravity, which is around 9.81 metres per second per second. That makes the force downwards from the weight of the water 0.0324 kilogram metres per second per second, more simply known as 0.0324 newtons (the unit of force named after the great man).

Acting upwards is the air pressure. At sea level this is around 101,325 newtons per square metre or 0.101325 newtons per square millimetre. The area it's acting on here is the cross-section of the straw. That's a disc around 3 millimetres in radius, so working out the area with πr^2 (where r is the radius of the disc), it's around 28.27 square millimetres. That gives an upward force on the water of 28.27 × 0.101325 = 2.86 newtons. The air pressure wins by a factor of 88 – that water is going to stay in place as long as it's a simple battle between gravity and air pressure.

When you get the water running through the straw from one glass to the other, you've made yourself a siphon. Once you do this, the balance of forces has changed. As before, we have the force of gravity downwards and the air pressure upwards on the bottom of the straw, but now the top of the straw is open. This means that the water in the straw feels some of the pressure coming from the air pushing down on the water in the top glass.

The water in the straw starts to drop. If a gap were to form it would be at a reduced pressure compared with the air pressure on the top of the glass that is feeding the straw. The combination of gravity and air pressure forces the water over the bendy bit of the straw and down the far end. It will continue to do so as long as there's water in the full glass covering the short end of the straw.

In a traditional domestic toilet there's an S-bend at the back, acting like the bendy bit of the straw. When you flush the toilet, a large rush of water enters the bowl and the air pressure on the bowl causes it to siphon, sucking material over the bend and away down the drains. Such an approach isn't ideal for an aircraft toilet, partly because it means carrying a large amount of heavy liquid to do the flushing – and weight means money in the aviation business – and partly because the more fluid there is, the more chance it will leak and cause technical problems.

The vacuum solution

Older aircraft toilets were similar to the chemical toilets used at festivals. A relatively small flush of (usually blue) chemicals was helped on its way by an electrical pump. However, as anyone who has attended a festival knows, such toilets tend to be smelly and easy to block, while still needing a storage tank of chemicals with the inevitable risk of leakage. Because of this, modern aircraft have vacuum toilets. Here, when the flush button is pressed, a vacuum is built up in a chamber behind the bowl. After a few seconds, with a distinctive sound, the vacuum is opened to the bowl, the pressure in the toilet drops, and the material in it is sucked out of the bowl, just as a vacuum cleaner sucks up rubbish.

Every now and then stories begin to circulate about passengers getting stuck onto the seat by the vacuum system. In 2002, BBC News carried a story that an American passenger had used the toilet on an SAS transatlantic flight and had pushed the flush button before she stood up. 'To her horror', trumpets the BBC, 'she realized that the powerful vacuum action had got her in its grip. Her body was sealed to the seat so firmly that it took airport technicians to free her.'

According to the report, the passenger was stuck in the Boeing 767's toilet for over two hours until the aircraft had landed and the cabin crew were able to call in technicians to help free her. The story even quotes an SAS spokeswoman as saying that the passenger would be compensated for her ordeal. 'She was stuck there for quite a long time', the spokeswoman commented.

It turned out on later examination by SAS that the whole thing was fictional. There was no record of such an incident, but stories like this were used in crew training to make sure that crew checked toilets, giving them guidance on what to do. The incident never happened. It's virtually impossible to do anyway, as the flush button is usually sited behind the toilet lid, so that to be able to flush it you have to be well away from the seat. If it were possible, it's unlikely that the seal would be good enough to trap someone for any length of time, especially as the vacuum system cuts off after a few seconds – but there's a significant risk of causing internal injury, so it's certainly not something to experiment with.

Meeting the night sky

Back in your seat, it may be getting dark (or maybe you took off in the dark). It might seem that flying in darkness reduces the science that you can check out from your airplane seat, but instead it opens up a whole new range of prospects.

Up above the clouds, the night-time sky is always clear, and won't suffer from the sky glow of street lamps that haunts a city. Only the poor quality of the window gets in the way of some great night-time sky viewing. Early on, as twilight shifts into darkness, you may see just one very bright star, near the horizon. This is most probably Venus, the planet that for a long time was thought to be most like the Earth.

A view of Venus

Venus is certainly similar in size to Earth, and though significantly closer to the Sun than we are, it was thought for a long time that its cloud cover could make the surface habitable. It was quite a shock when the first probes reached Venus and were almost instantly destroyed by a hellish 480°C average surface temperature. Lead is a liquid at this temperature. To make matters worse, the atmosphere, practically all carbon dioxide, is much thicker than our own, with a pressure on the ground of around 90 times that on Earth.

We only ever see Venus at dusk and dawn, which is how it got the nickname of the evening star. This is because it's closer to the Sun than we are, so it will always appear relatively close to the Sun in the sky. It's not usually bright enough to be seen in the day (though you can occasionally make it out), so it tends to appear low down as it follows the Sun below the horizon.

Although you can't make it out with the naked eye, even basic binoculars will show you that Venus, like the Moon (and for the same reason as the Moon – see below), goes through phases, so sometimes it's a full circle, while at others it's just a crescent. Venus is the brightest of the planets, followed by Jupiter and then Mars, both of which are outside our orbit, so aren't forced to stick close to the Sun in the sky. If you see another very bright 'star' it's probably one of these two – Mars is easily distinguished by its red tint.

There's not a huge amount of difference visually between these planets and the stars, except that the stars twinkle rather more, and the planets are noticeably brighter. Planets twinkle less because they're much closer, so less of a tiny point source of light, which means that they aren't so influenced by dust and heat ripples in the atmosphere. In absolute terms, they are much dimmer than the stars. Planets shine only by reflecting sunlight, whereas the stars are suns in their own right – but are so much further away that they appear faint. The nearest star is around four light years away. (A light year is a measure of distance, not of time. It's the distance light travels in one year.) Given that light travels 300,000 kilometres per second, that puts that star, Proxima Centauri, at a distance of around 9.5 trillion kilometres from the Earth.

The amazing Moon

The Moon is, of course, the brightest thing in the night-time sky. It's entirely coincidental that our natural satellite happens to look pretty much the same size as the Sun. This is because although the Moon is around 400 times smaller than the Sun, it's also around 400 times closer to us. We can see how close the size comparison is during a solar eclipse, when the Moon comes between the Earth and the Sun. Depending on the distance between the Earth and the Sun, which varies as our planet moves around its orbit, the Moon will sometimes entirely cover the Sun, but sometimes it will leave a ring of light around the outside, a so-called annular eclipse. So with just the right distance, the two are pretty well exactly the same visible size.

34. In an annular solar eclipse, a ring of sunlight is visible around the Moon.

This coincidental similarity will change over time, as the Moon's orbit is very gradually getting larger, making it appear smaller. In effect, the Moon's gravitational effect is slowing the Earth down. Because angular momentum of a system is conserved (remember the idea of an ice skater spinning faster when she pulls her arms in), the Moon gains momentum as the Earth loses it. This is a tiny effect – but a noticeable one. The Moon gets around 4cm further away from us each year.

The size of the Moon isn't always easy to judge. We've all seen the Moon look much bigger than usual (though not as ridiculously large as Hollywood often portrays it). This is a psychological effect, rather than a physical one. The visible size doesn't really change, but as we'll see later on, the picture we have of the world as seen through our eyes is a highly artificial construct. The best guess for why the Moon varies in size is that we tend to think it's bigger when it appears to be near physical objects that we know the size of: when the Moon is near the line of sight of a building or trees, for example. Our brain knows these things are relatively near, the theory goes, so it assumes that the Moon is a lot closer too.

We certainly are quite good at linking things that are really hugely distant from each other. When we look at the stars, we imagine that they form constellations – shapes that connect these points of light into a skeletal picture – where in reality there's no link between them. You only have to look at the Centaurus constellation in the southern sky – its brightest star, Alpha Centauri, is one of the nearest to ours, not much more than four light years distant. The next brightest in the constellation,

Beta Centauri (or Agena) lies 190 light years away, more than 45 times more distant. We are mistakenly linking together two objects that are separated by around 1,797,552,000,000,000 kilometres.

Our own star is much closer to Alpha Centauri than Agena is, yet we would hardly think of the Sun and Alpha Centauri as forming a pattern. Alpha and Beta Centauri are no more connected than Houston and Cairo are, simply because they lie near the same latitude. Our eyes and brains, looking for structure in the myriad of winking points in the sky, deceive us into finding patterns.

Experiment – How big is the Moon?

Here's a chance to see whether your brain is fooling you on the visible size of the Moon. Without checking (if the Moon is visible), guess what size coin, held at arm's length, would be the same size as the full Moon.

If there's a fairly full Moon in sight (or next time you can see one), try it out.

You will find that no coin is as small as the visible size of the Moon. A much better approximation to the size is the hole in a piece of hole-punched paper, held at arm's length. It really is that small, but our brains fool us. This is why when you take a photograph including the Moon it's usually so disappointing. On the photograph, our brains aren't fooled.

You can check this apparent size out with our distance estimation technique. The Moon is around 3,500 kilometres in diameter and around 380,000

---

kilometres away. So to find its apparent size in millimetres at 0.75 metres' distance from the eye, we need
to change the diameter of the Moon to millimetres
and the distance to metres, then multiply by 0.75. So
the Moon's apparent diameter at arm's length is 0.75
× 3,500 × 1,000 × 1,000/380,000 × 1,000 = 6.9mm. As
punch holes are around 5mm across, this is not a bad
approximation.

The changing face of the man in the Moon

If the Moon is visible out of the window at the moment,
it may well have a crescent shape. This effect arises from
the relative positioning of the Moon and the Sun. First
think about just the Moon and the Earth. At any one time,
around half the Earth should be able to see the Moon,
but in the daytime it's so relatively weak compared with
the Sun (moonlight is around 300,000 times weaker than
sunlight) that we often can't spot it.

The light of the Moon is purely reflected sunlight. The
Moon itself is not a great reflector – on average its coloration is a pretty dark grey – but the light from the Sun is
so powerful, and the night sky otherwise so dark, that it
seems very bright. This reflected source of the light is the
reason that we see the different shapes or phases of the
Moon.

Experiment – Phases of the Moon

It's easiest to envisage how the Moon's phases arise
by using objects. The ideal is to go into a dark room,

use a torch as the Sun and hold a tennis ball as the Moon, rotating it around your body (you represent the Earth). This isn't practical on the plane, but it's possible to get the idea using three objects – for example cups, bottles or glasses – on the table in front of you.

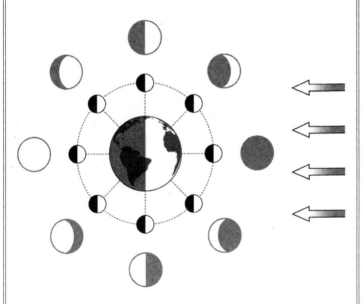

35. The phases of the Moon in its orbit around the Earth.

Place one object near the back of the table as the Sun. Position a second object near your edge of the table as the Earth. Now rotate a third object (the Moon) in an orbit around the Earth. As it passes around, imagine the light hitting it from the Sun, and how much of the lit-up bit of the Moon you could see from the Earth. When your Moon is directly between you and the Sun, the lit-up bit will be round the back, pretty well invisible – it will be a dark 'new moon'. (Half the Moon is still lit up, but it's the opposite side, which is often inaccurately called the dark side of the Moon.) As your Moon comes around, you will see first a small edge of it lit until the whole face is lit – the full moon. Then the lit part will shrink again until it has made a complete circuit.

You might wonder why there isn't an eclipse of the Sun every new moon and an eclipse of the Moon (where the Earth's shadow falls on the Moon) every full moon. This is because the Moon doesn't move around the Earth in an orbit that lines up with a line connecting the Earth to the Sun. As the Moon is off to one side, we usually see, for example, the full moon when the Moon is on the far side of the Earth from the Sun.

Lunar eclipses, where Earth's shadow falls on the Moon, happen much more often than solar eclipses (because the shadow of the Earth is much bigger with respect to the Moon than the shadow of the Moon is with respect to the Earth). An eclipse of the Moon is visible from half the planet – everywhere it's night – unlike a solar eclipse, which can be seen only in a narrow corridor on the Earth's surface. And lunar eclipses happen much more frequently – typically twice a year. During a lunar eclipse, the Moon doesn't go entirely dark, but instead takes on a strong red coloration, as light from the Sun passes through the Earth's atmosphere and is bent, just as light is bent passing through a prism. Less of the red light is scattered by the atmosphere, just as happens to make the sky blue, so the Moon appears red.

Welcome to the galaxy

It's just possible that from your window you will be able to make out something that's increasingly difficult to see from the ground as it becomes compromised by the glow of streetlamps – the Milky Way. This is a faint, arching band of glowing light that crosses much of the span of the sky. The Milky Way is what we can see of our home galaxy. You can imagine the whole galaxy as like a disc with a bulge towards the middle – if you saw it from the top it would look more like a spiral, but we're seeing it sideways on from our position in one of the spiral arms, and the visible part of this sideways-on disc forms the Milky Way.

The Milky Way is around 100,000 light years across, and around 1,000 light years thick, with at least 100 billion stars crammed into it, and quite possibly two or three times that number. (As you might imagine, counting stars on this scale isn't practical, so we have to rely on an estimate.) Although it's on a vast scale, it's sobering (somehow more so when you look out into the darkness from the tiny island of light that is a plane) to think that there are estimated to be at least 150 billion galaxies in the part of the universe we can see. There may be many more out there, but because light travels at a finite speed, and the universe hasn't been around for ever, we can see only so far.

The street light fantasia

If you arrive at your destination during the night, one of the more spectacular sights is likely to be the one that greets you as you descend towards the ground. Street lights may create havoc when viewing the stars, but in their own right these mundane bits of street furniture can create dramatic and beautiful vistas from the air.

Experiment – Illuminating head count

As a very rough rule of thumb, in towns and cities in industrialized countries there's one street light for every ten to twenty people. As you descend over a town, select a smallish square area and estimate how many of this sized square would fit in the whole illuminated mass of the town. Then take a rough count of the number of street lights in the area and multiply by ten for a lower estimate of population, doubling that for a higher estimate.

So, for instance, if your square would fit about ten times in one direction over the town and fifteen times in the other, that's 150 squares to cover the town. Let's say you counted 50 street lights in the square. That makes 7,500 lights in total. So the population is likely to be between 75,000 and 150,000.

The amazing eye

By now you're getting lower, and if it's daytime you may be able to see some of the features mentioned earlier in the book. If it's night, the lights in the cabin will be dimmed. This is a precaution in case you have to get out of the aircraft quickly, to ensure that your eyes have acclimatized to the low light conditions. The human eye is a remarkably flexible instrument. On a really clear night, it can see a candle flame 16 kilometres (10 miles) away. If it were dark enough, with clear windows, you could easily see a candle on the ground from a plane's cruising height. It takes only five or six photons of light to trigger a response. Sadly, thanks to air pollution and light pollution, there are relatively few places on Earth where you can make use of this capability.

The eye's mechanism is quite amazing. We have four different kinds of sensor in the eye, one type (rods) just handling black and white. There are about 120 million of these rods, which are significantly more sensitive than the three types of cones, 7 million in all, that deal with colour. When the light levels are low, the cones switch off, leaving our vision as shades of grey. (If you don't believe this, try it when you get home. Make a room really dark, then allow in the smallest amount of light – you won't be able to tell what colour objects in the room are.)

The colour-detecting cones are mostly bunched together near the centre of the eye's field of view. If light is weak, you can see things better if you don't look directly at them. That way the extra rods towards the edges of your vision can come into play. These are thought to be there to help us spot predators, creeping up on us in the dark.

When enough photons are captured by a rod or cone, each triggers a tiny signal. These signals are pre-processed in the eye – there are fewer connections in the optic nerve, which runs from the eye to the brain, than there are rods and cones. The combined signals pass up the optic nerve and stimulate areas of the brain. Most of the signals from the left eye go to the right side of the brain, and from right eye to left side, though some don't do this switch-over, to help your brain process 3D vision.

Making up a picture of the world

The brain has, in effect, a series of modules that handle shapes, edges, shading and other aspects of the visual nature of things you can see. From the information these natural processors generate, the brain builds an artificial picture of the outside world. It's important to realize that what we see really is artificial. This is why optical illusions work. When you look at one of these, your brain's mechanism for making up a picture of the world has been fooled. The eye/brain combination isn't like a camera, which simply records what's hitting the sensors – the brain makes things up where necessary.

If this weren't the case, we would have some problems. Where the optic nerve joins the back of the eye, there's a blind spot, leaving an area of sight where you simply can't see. The brain fills in what's happening in that area by educated guesswork. Similarly, your eyes are constantly jerking around in tiny, very quick movements called saccades. (This is partly how the brain covers the blind spot.) These are ironed out by the brain, leaving a steady view.

All this manipulation by the brain is why you can watch a movie on that seatback screen. TV and cinema works by projecting a series of still pictures. For a long time, the reason we saw these pictures as moving was thought to be due to 'persistence of vision' (you'll still see this explanation in many books and websites). This is rubbish. Not only is the eye too slow in responding to build up an image this way, if the still pictures did persist, the result would be a mess, not movement. Instead, it's the way the eye constructs a fake view using these different modules that allows it to be misled by a series of still images and think it's seeing movement.

Eyes wide

When the lights in the cabin are turned down, the eye is given a chance to respond to low light conditions. The iris of the eye – the coloured bit – has two sets of muscles. The sphincter muscles, the ones that pull the dark pupil in the centre smaller, loosen; while the dilator muscles, which pull the edges of the pupil open, contract. The result is that the hole allowing light into the eye becomes bigger. This can take a second or two before it reaches maximum aperture.

At the same time, the processing structures in the brain recalibrate to handle weaker light. One of the ways our brains fool us is by smoothing out variations in light intensity. If you go out early in the morning or at dusk, you may find that sensor-activated security lights come on, even though the light levels seem fine – this is because your brain/eye combination is compensating and not letting you realize just how low the light levels are.

Similarly, if you go from an electrically lit room inside to the outside on a bright sunny day, it doesn't seem hugely brighter, perhaps only a factor of two or three. In reality, outdoor sunlight can be 100 times brighter than typical indoor lighting. This is why, when using a video camera, the view will flare to practical invisibility when you move outside, before the electronics can compensate. The camera isn't as clever as the brain/eye combo at dealing with changes in light levels.

Remember that full moonlight is 300,000 times weaker than sunlight. If it's fairly dark outside, your eyes need all the help they can get in adapting to outside conditions, and dimming the cabin lighting makes that transition easier.

First touch on the runway

The plane lands with a puff of smoke from the wheels as friction takes a thin sliver off the surface of the tyres.

Friction is something we tend to ignore in the physics of motion – but in real life we ignore it at our peril. Friction has its good side. Imagine there was no friction. You couldn't pick anything up – it would just slide out of your grip. Think of the slippiest bar of soap, magnified many times over in difficulty of getting hold of it. Although it should in principle be easy to push objects along the ground without friction, you would have nothing to hold your feet in place – the action and reaction in Newton's third law would mean that you shot backwards whenever you tried to push something forwards.

In practice, of course, there *is* friction. In part this can be physical, when small irregularities in one surface fit into bumps and dips in another, like gears intermeshing. But most friction is electromagnetic. As we've seen, the atoms that make up objects consist of a positively-charged nucleus and negative electrons, forming a cloud of charge around the nucleus. These charges interact whenever two objects are brought together. That's how you manage to sit in your seat on the plane.

Think about it – 'solid' objects are mostly empty space. They consist of tiny atoms with big gaps between them – and the atoms themselves are mostly empty space. The positively-charged nucleus at the centre of an atom is absolutely tiny compared with the atom as a whole. Its relative size is like that of a fly in a building the size of the Albert Hall. The outer part of the atom does contain electrons, but these are tinier still, always moving, existing in a cloud of probability around the nucleus. If it weren't for the electrical charge on the parts of the atom, there's really no reason why you wouldn't sink straight through your seat.

In practice what happens is that the negative electrons on the outside of the seat repel the negative electrons on the outside of your body. You don't actually come into contact with the seat, you float just above it on a cloud of repulsion.

If that were all that was happening, you would be pretty well frictionless. But where there's a large electron cloud on one atom it can push the electrons on another atom out of the way, leaving more of a positive charge for it to cling on to. It's this electromagnetic attraction that's responsible for much of friction, rather like the way the little pieces of paper clung to the pen or comb in the static electricity experiment.

To overcome the 'stickiness' of friction we have to put energy into whatever it is that we're trying to move – and that energy is mostly converted to heat. You can feel this when you rub your hands together vigorously. When an aircraft's tyres hit the tarmac, they gain energy both from the collision and from friction on the surface. This energy ends up mostly as heat, vaporizing a thin outer layer of the tyre and producing that typical little puff of smoke.

Final steps

The journey has come to an end. Hopefully reading this book has made it seem shorter – and more interesting. It's easy to dismiss science as 'just nerdy stuff', but it really should be exciting. It's our best understanding of how the universe works, whether at the level of stars and galaxies or the everyday behaviour of the objects around us.

There are few chances to observe so much science in action as you get from your seat on a plane. Combining what you see out of the window with what's happening around you, everything from quantum physics and relativity to river formation and the workings of the eye have come into play.

With science as your guide, the everyday will never appear quite so ordinary again.

Picture credits

The author and publisher would like to thank the following for their permission to use copyright photographs:

29.   Air Canada (www.aircanada.com)
31.   Brandrodungswanderfeldhackbau (http://commons.
        wikimedia.org/wiki/File:787-flight-deck.jpg)
33.   Pipimaru (http://commons.wikimedia.org/wiki/
        File:Himalayan_mountains_from_air_001.jpg)
34.   A013231 (http://commons.wikimedia.org/wiki/
        File:Solar_annular_eclipse_of_January_15,_2010_in_
        Jinan,Republic_of_China.JPG)

All line drawings produced by Nick Halliday (info@
hallidaybooks.com).

Index